NATURE
LOVES TO HIDE
Quantum Physics and the Nature of Reality,
a Western Perspective

Revised Edition

NATURE LOVES TO HIDE

Quantum Physics and the Nature of Reality, a Western Perspective

Revised Edition

Shimon Malin

Colgate University, USA

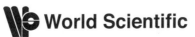

World Scientific

NEW JERSEY · LONDON · SINGAPORE · BEIJING · SHANGHAI · HONG KONG · TAIPEI · CHENNAI

Published by

World Scientific Publishing Co. Pte. Ltd.

5 Toh Tuck Link, Singapore 596224

USA office: 27 Warren Street, Suite 401-402, Hackensack, NJ 07601

UK office: 57 Shelton Street, Covent Garden, London WC2H 9HE

British Library Cataloguing-in-Publication Data
A catalogue record for this book is available from the British Library.

Cover image: Bubble chamber: Omega production and decay. Copyright (1973) CERN.

Cover photo: Niels Bohr and Albert Einstein. Photo by Paul Ehrenfest. Source: Wikipedia.

ISBN-13 978-981-4324-56-4
ISBN-10 981-4324-56-6
ISBN-13 978-981-4324-57-1 (pbk)
ISBN-10 981-4324-57-4 (pbk)

Printed in Singapore by B & Jo Enterprise Pte Ltd

This book is dedicated to the memory of
William C. Segal,
teacher and friend

CONTENTS

ACKNOWLEDGMENTS

I wish to express my gratitude to the many friends, colleagues, and members of my family who helped me in this project. First and foremost, I wish to express my thanks to William C. Segal, to whom this book is dedicated. The beneficent influence of my long acquaintance with him, and the perspective on life that he shared with me during this long acquaintance, permeate this book. In particular, the views expressed in Chapter 18 on our place in the universe reflect his teaching. A full articulation of this teaching can be found in his book *Opening: Collected Writings of William Segal, 1985–1997*.

I wish to acknowledge a debt of gratitude to Anthony Damiani for introducing me to the teachings of Plotinus. It is a pleasure to thank my wife, Tova, and my sons, Nadav and Yonatan, for their comments and criticisms; special thanks to my daughter, Daniella, for her work on the illustrations as well as for her critical comments. I thank Jane Ehrlich for bringing the illustrations to their final form. I am grateful to Professor Abner Shimony for his encouragement and comments and for the enlightening discussions we had. I had many enlightening discussions about Plato, Plotinus, and Whitehead with Prof. Anne Freire-Ashbaugh as well. Thanks are due to Kirk Jensen of Oxford University Press for his steadfast support and useful suggestions and to India Cooper for excellent copy editing. And I wish to thank the following friends and colleagues for their critical reading of the manuscript: Anthony Aveni, Jesse Carr, Jane Collister, Patricia Glazebrook, Martha Heyneman, John McVeigh, S. Chandu Ravela, Cynthia Reeves, Naomi Schwarz, Jonathan Swinchett, Frank Taplin Jr., and Richard Thompson. Finally, I wish to thank Gabriel Greenberg for his help with the illustrations, Roger Williams for the photograph of the double-slit experiment, and Frank Smith III for his editorial help.

INTRODUCTION

Decades ago, when I was a junior in college, I took a course in quantum mechanics and fell in love with it. I spent long nights trying to decipher Paul Dirac's book *The Principles of Quantum Mechanics*, sometimes spending hours over a single page. I marveled at the depth of insights it contained and felt a great mystery there. Although I gained a working knowledge of quantum mechanics rather quickly, I felt that only a small part of the mystery was accessible to me. This was the beginning of a lifelong engagement. For the next forty years I continued to investigate this mystery. The results of my research were published in professional journals and reported in international conferences. Until now, however, I limited my communications to professional circles.

This book was written for a general audience. Its theme is the insights provided by the quantum theory about the nature of reality: If, as Heraclitus said twenty-five centuries ago, "Nature loves to hide," what secrets can quantum mechanics help us uncover?

"Nature loves to hide." This statement resonates in us, as it did in the ancient Greeks. Behind the display of phenomena there is a hidden reality. But what is this hidden reality? What is its relationship with the sensory world? Do we have a world-view that can encompass both the hidden and manifest aspects of nature?

Seven hundred years ago these questions would have been answered by Dante's vision. It provided the Western mind with a comprehensive world-view, integrating the various aspects of the universe into one magnificent whole, a whole in which we humans played an organic and central role. One hundred years ago these questions would have been answered by a Newtonian "clockwork universe," a model of a universe which is completely mechanical. Everything that happens has been predetermined by the laws of nature and by the state of the universe in the distant past. The freedom one feels in regard to one's actions, even in regard to the movements of one's body, is an illusion.

The world-view expressed by Dante, as well as the Newtonian one, is complete and coherent. There were times in the history of Western civilization, however, when no coherent and comprehensive world-view was in existence, times when the old world-view had collapsed and a new one had not yet been found.

Dante's cosmology collapsed in the middle of the sixteenth century with the advent of the Copernican revolution. The Copernican revolution was troubling for the Western mind not only because it demolished Dante's paradigm but also because it did not provide an alternative. Copernicus discovered a new astronomical theory, not a new and comprehensive world-view. The period that started in 1543, with the publication of Copernicus's *De Revolutionibus Orbium Coelestium*, and ended in 1687, with the publication of Newton's *Principia Mathematica Philosophiae Naturalis*, was a time of transition, a time in which Western civilization was devoid of a coherent set of beliefs regarding the nature of reality.

Similarly, at the present time, we live, as Thomas Berry said, "between stories." With the advent of Einstein's Special and General Theories of Relativity the Newtonian ideas about space and time were thrown overboard. And with the advent of the quantum theory the Newtonian concept of matter was replaced by a radically different concept. These theories provided physics with new paradigms; they did not, however, provide Western civilization with a comprehensive world-view.

The human mind abhors a vacuum. When an explicit, coherent world-view is absent, it functions on the basis of a tacit one. A tacit world-view is not subject to a critical evaluation; it can easily harbor inconsistencies. And, indeed, our tacit set of beliefs about the nature of reality is made of contradictory bits and pieces. The dominant component is a leftover from another period, the Newtonian "clockwork universe." We cling to this old and tired model because we know of nothing else that can take its place. Our condition is the condition of a culture that is in the throes of a paradigm shift.

The term "paradigm shift" is a misnomer. As I read it I see, in my mind's eye, an image of a gentle movement; I see myself lounging lazily in a sailboat, as the wind shifts, almost imperceptibly, from west to northwest. In fact, a paradigm shift concerning the nature of reality is an earth-shaking affair. It involves a profound crisis for the individual and major upheavals for society. Much is at stake. In the case of the present paradigm shift the fate of the planet itself has been called into question.

A major paradigm shift is complex and difficult because a paradigm holds us captive: We see reality *through* it, as through colored glasses, but we don't know that; we are convinced that we see reality as it is. Hence the appearance of a new and different paradigm is often incomprehensible. To someone raised believing that the earth is flat, the suggestion that the earth is spherical would seem preposterous: If the earth were spherical, wouldn't the poor antipodes fall "down" into the sky?

And yet, as we face a new millennium, we are forced to face this challenge. The fate of the planet *is* in question, and it was brought to its present precarious condition largely because of our trust in the Newtonian paradigm.

The Newtonian world-view has to go; and, if one looks carefully, the main features of the new, emergent paradigm can be discerned. The search for these features is what this book is about. We will find them by following four different lines of exploration.

The first line of exploration is based on a set of hints that are provided by quantum mechanics. Although quantum mechanics is not a comprehensive world-view, it is replete with suggestions. The emergent paradigm will have to encompass quantum mechanics within its ken. The "weird" aspects of the quantum theory provide us, therefore, with fertile ground for our search; the feeling of weirdness indicates that the theory is inconsistent with the prevailing world-view. This feeling should disappear once it is replaced by the new one. If one believes that the earth is flat, the story of Magellan's travels is puzzling: How is it possible for a ship to travel due west and, without changing direction, arrive at its place of departure? Obviously, when the flat-earth paradigm is replaced by the belief that the earth is spherical, the puzzle is instantly resolved.

The founding fathers of quantum mechanics were well aware of the earth-shaking implications of their discoveries. When I started reading the philosophical essays of Albert Einstein, Niels Bohr, Werner Heisenberg, and Erwin Schrödinger, I was impressed by the depth of their thoughts and surprised to discover how often and how passionately they disagreed with each other. These philosophical writings and illuminating controversies are the second line of exploration. Heisenberg will introduce us to a new and revolutionary view of the nature of atoms and subatomic particles; we will explore, with Schrödinger, "the principle of objectivation," a generally unacknowledged yet fundamental limitation of science as it is currently practiced; and we will follow closely the celebrated life-long controversy between Einstein and Bohr. The clash of these two great minds did much to clarify the message of the quantum theory.

I found the essays of the founders of quantum mechanics deeply engaging but incomplete. Einstein, Bohr, Schrödinger, and Heisenberg were physicists, not philosophers. None of them attempted to construct a philosophical system. I felt, however, that the mystery at the heart of the quantum theory called for a revolution in philosophical outlook. This lacuna was filled when I encountered the writings of Alfred North Whitehead. During the 1920s, when quantum mechanics reached maturity, Whitehead constructed a full-blooded philosophical system that was based not only on science but on non-scientific modes of knowledge as well. An examination of Whitehead's thought and its relationship to quantum mechanics is our third line of exploration.

The influence of a fading paradigm goes well beyond its explicit claims. All paradigms include subterranean realms of tacit assumptions, the influence of which outlasts the adherence to the paradigm itself. Copernicus, for example, removed the earth from the center of the universe but continued to believe that the laws of nature in the heavens are different from those on earth. In our case, the belief that science, and science alone, holds the key to the understanding of the nature of reality is just such an assumption. We believe, as the scientists and philosophers of the Enlightenment did, that when we wish to find out the truth about the universe, non-scientific modes of processing human experiences can be ignored. Poetry, literature, art, music are all wonderful; but, in relation to the quest for knowledge of the universe, they are irrelevant. It was Alfred North Whitehead who pointed out the fallacy of this assumption. In this, as well as in other aspects of his thinking, Whitehead was decades ahead of his time.

According to Whitehead, the building blocks of reality are not material atoms but "throbs of experience." Whitehead offers a shift from a mechanistic paradigm to an organic one, "a universe of experience." Whitehead formulated his system in the late 1920s, and yet, as far as I know, the founders of quantum mechanics were unaware of it. It was not until 1963 that J. M. Burgers pointed out that Whitehead's philosophy accounts very well for the main features of quantum mechanics, especially the "weird" ones.

Heisenberg was the first to draw attention to the fact that the concept of matter that emerges out of quantum mechanics, a concept which is very different from the one ordinarily associated with the word "matter," is largely a Platonic one. Not surprisingly, Whitehead's system is Platonic in its tenor. The last three chapters of the book are devoted to the fourth and last line of exploration. We will supplement our study of Whitehead's philosophy by delving into its Platonic origins and, in particular, into the cosmological thought of Plotinus, the great neoplatonist of the third century A.D.

This line of exploration will enable us to look into essential questions that lie outside the Whiteheadian system: Are there different levels of being, i.e., are some aspects of reality "higher" or "deeper" than others, and if so, what is the structure of the hierarchy? What is our own place in the universe? And, finally, what is the relationship between the great aspiration within physics, the quest for "a theory of everything," one theory that accounts for all the known facts, and the deepest aspirations of the human spirit?

Dante's vision provided each human being with a place of great significance and great dignity. The author/pilgrim ends his journey in the highest celestial spheres. With Copernicus we began our downward spiral. It led to the current cosmological world-view, according to which we are insignificant inhabitants of an insignificant speck of dust in a universe of swirling galaxies. An attempt to endow us with a cosmological meaning in such a universe seems absurd. And yet, *this very universe is just a paradigm, not the truth.* When you reach the end of the book you may be willing to join me in entertaining an alternative view according to which, surprisingly, the dignity that Dante endowed us with is restored, albeit in a post-postmodern context.

The philosophical implications of quantum mechanics have been the subject of a number of recent books that tend to emphasize connections between quantum physics and Eastern religions. I believe that investigations of such connections are important. In the present book, however, we are concerned exclusively with the Western tradition. We explore the relationships between quantum mechanics and Western philosophical thinking, from Plato and Plotinus to Whitehead and the quantum physicists.

Some aspects of the interpretation of quantum mechanics presented here express the consensus of the physics community. Other aspects are shared by some and objected to (sometimes vehemently so) by others. Still other aspects express my own views and convictions. I have tried to make it clear, as we go along, which is which.

The experience of writing this book turned out to be more difficult than I anticipated, and also more rewarding. When I came across the challenge of explaining particularly difficult ideas, I discovered that a conversational mode would be helpful. Therefore I introduced two fictional characters, Julie and Peter; they started out as college friends, studying psychology, and later became astronauts. Their conversations with each other and with me were fun to write. I hope that they will be not only illuminating but also fun to read.

PART ONE

The Quandary

In Part I we introduce the main ideas of quantum mechanics against the back-drop of the celebrated Bohr-Einstein debate.

Albert Einstein was firmly committed to a world-view based on three tenets: *realism*, the belief that the physical world consists of objects which exist "from their own side," i.e., independently of consciousness; *locality*, the assumption that an event in one place can affect an event in another place only if there is enough time for a signal traveling no faster than the speed of light to propagate from one to the other; and *determinism*, the belief that every present or future event can be fully accounted for as the effect of past causes. This last tenet was encapsulated in Einstein's phrase "God does not play dice."

The combination of the first two tenets is called "local realism." If quantum mechanics is a complete, fundamental theory of nature, it violates both local realism and determinism. Einstein believed, therefore, that quantum mechanics is not a complete, fundamental theory of nature.

Niels Bohr was committed to a radically different point of view. He firmly believed that quantum mechanics was complete, and that one should accept its implications, wherever they lead. The quantum theory, as it turned out, led Bohr to far-reaching conclusions, such as his "framework of complementarity": To completely describe a quantum system, one needs not one model, but two. An electron, for example, is modeled as a wave and also as a particle. Such models are complementary rather than contradictory because they apply to different conditions.

The Bohr-Einstein debate lasted three decades and was unresolved at the time of their deaths. The controversy seemed to be a philosophical one, a matter of point of view. Quite unexpectedly John Bell proved, in 1964, that the debate can be settled by an experiment. He proposed a "thought experiment"—an experiment one thinks about rather than performs—which showed that, regardless of whether quantum mechanics is or is not complete, it violates local realism. A few years later it became feasible to actually carry out the experiment. When it was performed, it proved that *nature violates local realism*.

This leads us to the crux of our quandary. Local realism is an eminently reasonable world-view; it is, in fact, the world-view we hold, tacitly or explicitly. According to quantum mechanics, and according to Bell's experiment, it has to be given up. What will take its place?

1. Mach's Shadow

During the first conversation between Albert Einstein and Werner Heisenberg, Einstein was critical of Heisenberg's "new quantum mechanics." Heisenberg tried to defend the new theory: He pointed out that in its formulation he followed the requirements of Ernst Mach, the same requirements that guided Einstein in his discovery of Special Relativity. Einstein, however, did not budge. His astonishing comment "It is the theory which decides what we can observe" made a deep impression on young Heisenberg. This conversation was a precursor of the celebrated Bohr-Einstein debate.

~

All physicists of the last century saw in classical physics a firm and final foundation for all physics, yes, indeed, for all natural science. . . . It was Ernst Mach who, in his *History of Mechanics*, shook this dogmatic faith; this book exerted a profound influence upon me in this regard while I was a student.

—Albert Einstein

1. The New Quantum Mechanics

On April 28, 1926, a young lecturer from the University of Göttingen addressed the physics colloquium at the University of Berlin. The colloquium was a venerable institution, attended by the entire staff of the physics department, a department that was, at the time, the center of physics research in Germany and beyond.

The name of the speaker was Werner Heisenberg, and the audience he addressed was august indeed. It included Max Planck, who started the quantum revolution in 1900 with his discovery that light is emitted and absorbed in discrete packets of energy called "quanta," Max von Laue, who deciphered the nature of X-rays and revolutionized the study of crystals, Walter Nernst, who discovered the third law of thermodynamics, and Albert Einstein.

The subject of Heisenberg's lecture was "the new quantum mechanics," a theory that he developed in 1924 and 1925 in collaboration with Max Born and Pascal Jordan. The word *new* distinguishes the theory from what is now called "early quantum mechanics," a collection of theories formulated by Max Planck, Albert Einstein, Niels Bohr, and others during the first two decades of the twentieth century.

3

The new theory provided "a coherent mathematical framework, one that promised to embrace all the multifarious aspects of atomic physics."[1] Its discovery was an impressive achievement. And yet the presentation was bizarre. The theory was supposed to account for atomic phenomena, but Heisenberg was very careful not to describe what is going on inside atoms; in particular, he was careful not to address the burning issues of the day—"quantum jumps" and "the wave-particle duality."

The term "quantum jumps" refers to a model of the hydrogen atom that Niels Bohr proposed more than a decade earlier. According to Bohr's model, atoms are like miniature solar systems: The light electrons revolve around massive nuclei, just as the planets revolve around the sun. But there is a crucial difference: In the case of the solar system there is no special reason for the planets to be at particular distances from the sun. They just happen to occupy the orbits in which we find them; they might just as well have occupied other orbits. In Bohr's model, however, only certain orbits are "allowed." One never finds electrons anywhere except at these allowed orbits. But since they do change orbits, they must be capable of "jumping" from one allowed orbit to another. It is precisely during such jumps that an atom emits or absorbs packets, or "quanta" of light, also known as "photons." What is the nature of these quantum jumps? Should one think of an electron inside an atom as a little object circulating around a nucleus in an allowed orbit and, in some circumstances, jumping to another allowed orbit? If so, why don't we ever see an electron while it is jumping? This was one issue that Heisenberg was careful to avoid.

The other major issue that he ignored was "the wave-particle duality." In everyday life particles and waves are very different entities. A particle is an *object* that moves in space, while a wave is a *pattern of vibrations* that propagates from one place to another. The pattern is carried by a medium, but the particles that make up the medium do not go with the wave. Consider, for example, what happens when one drops a pebble into a quiet pond. A circular wave starts propagating, but the water particles do not move away from the pebble. They simply vibrate up and down, and these vibrations cause the circular pattern to travel further and further away from the place where the pebble was dropped.

In everyday life a phenomenon can be either a particle or a wave; it cannot be both. If an object travels, it is a particle. If a pattern travels while the medium in which it travels just vibrates, it is a wave. A tennis ball is a particle. Sound is a sequence of waves. As we will see, however, this clear distinction was put into question by the results of experiments on light and on electrons. In some circumstances light and electrons behave as if they were particles; in others, as if they were waves. What are they, really? This issue became known as "the wave-particle duality."

When Heisenberg presented his theory at the physics colloquium he was careful to avoid anything that seemed like a description or a presentation of a model. He confined himself to the presentation of a mathematical procedure for calculating the results of experiments. He was, incredible as this may sound, firmly convinced that in doing so he provided a full explanation of the atomic phenomena he was discussing! Young Heisenberg, like young Einstein before him, was thinking and working under the long shadow of Ernst Mach.

2. The Legacy of Ernst Mach

Nowadays the ideas of Ernst Mach are hardly known, even among intellectuals. His name is remembered only in the context of the term "Mach number," a unit of measurement of the speed of airplanes. The name of the unit was chosen to honor Mach's contribution to the study of aerodynamics; his contributions to philosophy are all but forgotten. At the turn of the century, however, Ernst Mach was a towering figure, influential not only in physics and philosophy but even in sociology and politics. No less a leader than Vladimir Ilych Lenin found it necessary, in 1908, to set aside his pressing duties as the head of the Bolshevik party and devote a few months to the writing of a voluminous book, *Materialism and Empiriocriticism*, devoted in large measure to the refutation of Mach's philosophy. Lenin saw the spread of Mach's views as a threat to Carl Marx's philosophy of "dialectical materialism," which was the theoretical foundation of the communist revolution he was preparing.

The forbidding term "empiriocriticism" is the name of Mach's philosophical system. The system must have been "in the air" in the latter part of the nineteenth century, because it was proclaimed, more or less simultaneously and independently, by two thinkers, Ernst Mach and Richard Avenarius. Empiriocriticism is both a philosophical system regarding the nature of reality and a philosophy of science. The following summary of some of its main ideas will give a taste of Mach's thought.

Science, according to Mach, is nothing more than a description of facts. And "facts" involve nothing more than sensations and the relationships among them. Sensations are the only real elements. All the other concepts are extra; they are merely imputed on the real, i.e., on the sensations, by us. Concepts like "matter" and "atom" are merely shorthand for collections of sensations; they do not denote anything that exists. The same holds for many other words, such as "body."

Mach carried his philosophy to its logical conclusion. Consider the case of a pencil that is partially submerged in water. It looks broken, but it is really straight, as we can verify by touching it. Not so, says Mach. The pencil in the water and the pencil out of the water are merely two different facts. The pencil in the water is really broken, as far as the fact of sight is concerned, and that's all there is to it.

Since science is, for Mach, just the description of facts, it does not aim at finding the truth about reality. It does not aim at finding the truth about anything. Its sole function is the achievement of "economy of thought," the description of the greatest possible number of facts using the smallest possible mental effort. A law of nature is valuable not because it is, in any sense, true but because it is a concise description of a large number of facts. Consider, for example, the phenomenon of free fall. One way to describe it is to create an enormous collection of data covering all the results of all the experiments conducted with falling bodies. Another way is to formulate the law of free fall, the law that says that the velocity of the falling object keeps increasing at a constant rate. According to Mach, the second way is superior to the first only because it is more economical. As far as "understanding" goes, both ways are equal.

Empiriocriticism arose as a reaction to the speculative German philosophy of the nineteenth century, an entangled, verbose mess of intricate "world-views," having little to do with either empirical evidence or clarity of thought. In this climate the simplicity, directness, and compelling logical coherence of Mach's presentations were a breath of fresh air. Many scientists were fascinated. Mach's approach cut through persistent dichotomies, e.g., matter vs. mind, and demanded an unprecedented rigor of thought. Every concept used in science had to have an "operational definition": One was not allowed to name a quantity unless one could specify how it could be measured. This led to fruitful reexamination of basic concepts, such as space, time, and energy. Yet this call for a precision of thought was deeply problematic. It was arrived at on the basis of faulty metaphysics, as we shall see.

Lenin was right, I believe, when he considered Mach a great physicist and a small philosopher. The blatant fallacies in Mach's philosophical arguments have been pointed out by many, Lenin and Einstein included. Einstein accepted Mach's system in his youth and disowned it in his forties. We will come to Einstein's thoughts on the matter in the next section. Here I will limit myself to a few critical comments of my own.

One role of science is to explain phenomena, and an explanation is different from "economy of thought." Consider the example of tides. People made accurate tables of the times of high and low tides in many locations, but the phenomenon of tides was not understood until Newton came along and explained it as the joint effect of the gravitational pull of the sun and the moon on the waters of the oceans. This discovery did not make it possible to calculate the times of high and low tides in specific locations. These depend on many complicated factors, such as the contours of the shores, the depth of the oceans at other locations, and so on; the complexity of these factors makes it impossible to calculate the tables of tides on the basis of Newton's laws. Newton's discovery did not lead to economy of thought; the tide tables continued to be produced from the records of local observations. But it did explain the phenomenon of tides.

Furthermore, the view of science as merely a system for thought economy is contrary to the experiences of many great scientists. They experience their acts of discovery as acts of seeing into the hidden workings of nature, not as acts of figuring out how to condense large bodies of information into "economical" packages. Heisenberg himself described his experience of discovering the new quantum mechanics in the following words:

At first, I was deeply alarmed. I had the feeling that, through the surface of atomic phenomena, I was looking at a strangely beautiful interior, and felt almost giddy at the thought that I now had to probe this wealth of mathematical structures that nature had so generously spread out before me.[2]

It seems to me that Mach's view is especially deficient in that it limits "the real" to the sensory. It implies that the attempts to explore the depths of reality are meaningless, because reality does not have any depth! And yet, sometimes, while listening to a piece of music, we feel that the sounds are merely a vehicle, and what they convey is something else, something that resonates in us on a level that is deeper than that of an ordinary sensation.

3. "It Is the Theory Which Decides What We Can Observe"

When Heisenberg's lecture was over, Einstein invited the young lecturer to accompany him on his walk home. The opportunity to have a conversation with Einstein must have been exciting for Heisenberg, and he must have expected to receive a pat on the back. After all, his achievement in atomic physics was accomplished using the same approach that guided Einstein two decades earlier when he discovered Special Relativity, the theory that revolutionized the concepts of space and time! Einstein arrived at his theory by following Mach's approach and analyzing very carefully the operational meaning of the statement "Two events that took place far away from each other happened at the same time." Similarly, Heisenberg arrived at the new quantum mechanics by analyzing very carefully what was actually being measured in atomic experiments. In doing that he was guided, in fact, by a comment of Einstein's, reported to him by a friend, to the effect that "physicists must consider none but observable magnitudes while trying to solve the atomic puzzle."[3]

To Heisenberg's surprise and dismay, however, Einstein did not like what he had heard at the colloquium. He criticized Heisenberg for following his own advice!

"You don't seriously believe," Einstein said, "that none but observable magnitudes go into a physical theory?"

"Isn't that precisely what you had done with relativity?" Heisenberg protested. "After all you did stress the fact that it is impermissible to speak of absolute time, simply because absolute time cannot be observed; that only clock readings, be it in the moving reference frame or the system at rest, are relevant to the determination of time."

"Possibly I did use this kind of reasoning," Einstein responded, "but it is nonsense all the same . . . on principle it is quite wrong to try founding a theory on observable magnitudes alone. In reality the very opposite happens. It is the theory which decides what we can observe."[4]

"It is the theory which decides what we can observe!" What did Einstein mean?

Figure 1.1 What do you see?

Look at Figure 1.1. What do you see? I showed it to two of my friends, Alan, a musician, and Jane, an electrical engineer. Alan said, "This is a bunch of lines. I have no idea what they mean." Jane got excited when she saw the picture. "This must be a photograph of a cloud chamber event," she said. "Most of these sluggish lines must represent slow electrons. But look at this line across the picture! This is a really fast one!"

Both Alan and Jane saw facts. But they saw two different sets of facts. Alan had no interest in physics; he did not know the theory behind the photographs, so what he observed was almost meaningless. Jane, who had studied physics, knew that a cloud chamber is a detector of atomic and subatomic particles, and she knew how it works: It is a transparent jar that contains supercooled gas, that is, a gas that has not condensed to become liquid, in spite of being cold enough to do so. Supercooled gas is unstable. Any slight disturbance, such as the passage of an electron, is enough to trigger condensation along the path of passage; the tiny drops of liquid that show up in the photograph indicate the trajectory of the electron. She was not thinking about that when she looked at the photograph, and yet it was this theory that determined what she observed.

And there is more. Alan too saw something, a picture that looked like a collection of lines. To see even that much, he too relied on theory. Both Alan and Jane know the principles of photography; and they understand seeing as a process that is initiated by light impinging on the retina and continues with signals traveling through the nervous system and reaching the brain. "Along this whole path," Einstein explained, "from the phenomenon to its fixation in our consciousness, we must be able to tell how nature functions, must know the natural laws at least in practical terms, before we can claim to have observed anything at all."[5] The understanding of photography and of seeing need not be explicit. I believe that this is the import of Einstein's phrase "at least in practical terms." Perhaps the word "theory" should have been replaced by the word "paradigm," which includes both explicit and tacit knowledge and belief. At any rate, there is no question that theory, i.e., explicit knowledge, and belief as to how nature works do each play an essential part in an act of observation.

If the theory decides what we can observe, how is it possible to come up with a new theory, then confirm it by observations? The theory that "determines what we can observe," Einstein explained, contains many elements. The discovery of a new theory involves changing one of these elements and keeping the others intact. For example, analyzing Figure 1.1, one can come up with a new theory about the electron; but the theories of supercooled gases, and of the principles of photography, stay the same. As we will see later on, Heisenberg made his second great leap, the discovery of the celebrated uncertainty principle, precisely through such a reinterpretation of what seemed like electron trajectories in a cloud chamber.

Listening to Einstein's explanations, Heisenberg was torn between his belief in Mach's doctrine, with which he was not prepared to part, and the realization that Einstein was right. "The idea that a good theory is no more than a condensation of observations in accordance with the principle of thought economy surely goes back to Mach," he said, "and it has, in fact, been said that your relativity theory makes decisive use of Machian concepts. But what you have just told me seems to indicate the very opposite. What am I to make of all this, or rather what do you yourself think about it?"[6]

Einstein responded by explaining that he was no longer an adherent of Mach's philosophy. He regarded Mach as being too naive. He suggested considering simple everyday concepts, like the word "ball." What does the word "ball" mean? Is it merely the result of thought economy, a condensation of the sensory experiences that involve balls? No, Einstein argued. *We are convinced that the ball really exists,* and this feeling of actual existence goes beyond thought economy. It means, for example, that we expect the ball to give us new experiences in the future; we associate with the word "ball" possibilities and expectations, which a mere condensation of past experiences does not include. Einstein saw Mach's ideas as naive, because Mach assumed that he knew what it means "to observe" and did not appreciate the complexity of an act of observation. He did not realize, for example, that "it is the theory which decides what we can observe."

4. Leaving Mach Behind

The conversation between Einstein and Heisenberg moved on to other subjects, such as the question of what can and cannot be described by an acceptable quantum theory. Einstein believed that the new quantum theory presented by Heisenberg was deficient, that it left too many issues unresolved. Heisenberg admitted that the theory was still too new to tackle the thorniest issues. They parted without an agreement, but with the expectation of continuing their discussions in the future.

The conversation with Einstein must have been, for Heisenberg, a disappointing experience. And yet Einstein's statement "It is the theory which decides what we can observe" made a strong impression on him and was later instrumental in leading him to the discovery of the uncertainty principle.

Unbeknownst to either of them, this conversation was the first move in what was to become the most remarkable controversy in twentieth-century physics. Because of the differences in age and stature in the scientific world (Heisenberg was twenty-five, fresh out of graduate school, while Einstein, at forty-seven, was at the height of his acclaim), Heisenberg was in no position to confront Einstein with a new, revolutionary paradigm about the task of physics, especially since the Machian position he was trying to defend was, indeed, indefensible. By the autumn of 1927, however, Einstein met his match.

Niels Bohr, the man who discovered the first successful model of the atom, was, more or less, a contemporary of Einstein and, just like Einstein, an acknowledged authority on the foundations of physics. The Solvay Conference in Brussels, in 1927, was the site of the first powerful clash between these two great thinkers, who, while having the highest respect for each other, totally disagreed about practically everything concerning the budding quantum theory. Their clash was a clash of paradigms. The disagreement about the quantum theory uncovered a basic disagreement about what physical reality is and what can be known about it.

In the discussions that took place between Einstein and Bohr, Mach was left behind. Both thinkers were too mature and too profound to take Mach seriously. By the 1920s Einstein had replaced his earlier adherence to Mach's philosophy with his own version of realism, a belief in an objective world, the existence of which is independent of acts of perception. Bohr came from an altogether different philosophical tradition. Influenced by Harald Høffding, who was, in turn, influenced by Søren Kierkegaard, Bohr was neither interested in Mach nor attached to realism. When it became clear that Einstein's version of realism was incompatible with quantum mechanics, Bohr made heroic yet vain efforts to move Einstein away from his cherished philosophical stance. Believing that quantum mechanics is a fundamental theory, he tried to usher Einstein into the new philosophical vistas that quantum mechanics opened up, but failed. As we go along we will explore these vistas, as well as Einstein's brilliant defense of his philosophical position. The presentation of this philosophical position is the subject of the next chapter.

2. Einstein's Dilemma

In this chapter we discuss Einstein's theory of Special Relativity and the three tenets of his world-view: realism, locality, and determinism. The belief in locality is based on Special Relativity, and the belief in realism is based on everyday experiences. But why did Einstein adhere to determinism so tenaciously? Uncovering the source of his belief that "God does not play dice" involves a detour into St. Augustine's response to the question "What is time?"

~

Time present and time past
Are both perhaps present in time future,
And time future contained in time past.
If all time is eternally present
All time is unredeemable.

—T. S. Eliot

1. "God Does Not Play Dice"

Einstein's conviction that quantum mechanics is not a fundamental theory of nature was a consequence of his paradigm, his firm beliefs about the nature of reality.

"It must have been around 1950," writes Einstein's biographer Abraham Pais. "I was accompanying Einstein on a walk from The Institute for Advanced Study [in Princeton] to his home, when he suddenly stopped, turned to me and asked me if I really believed that the moon exists only if I look at it."[1]

When he realized the fallacy of Mach's philosophy, Einstein became a realist. (Mach was not a realist. For him, the elements of reality, i.e., sensations, did not point to a "real" world beyond themselves.) It was obvious to Einstein that there is an objective world, whose existence is independent of acts of consciousness. He was not, however, a naive realist. He did not believe that the world we perceive is the world as it is. He appreciated Kant's finding that the world as we perceive it, the phenomenal world, is largely a creation of our own minds. He believed, however, that characteristics of the independently existing reality can be discovered through science and that scientific theories, or conceptual models, are our only access to reality.

In addition to being a realist, Einstein adhered to the tenet of locality: Nothing can propagate faster than the speed of light. As we will see, locality is a consequence of Special Relativity. The Einsteinian world-view was thus based on "local realism," the combination of realism and locality. In addition, Einstein was a firm believer in strict causality, also called "determinism," which claims that there is no randomness in nature, that the present and future states of the universe, down to the smallest detail, are the result of the state of the universe in the distant past. Einstein expressed this belief in his well-known phrase "God does not play dice." As we will see, strict causality is a claim that quantum mechanics denies. According to quantum mechanics the laws of nature are statistical; individual events in the present and the future are not entirely determined by the past.

Einstein's adherence to determinism made the gulf between his views and Bohr's views unbridgeable. For Bohr, quantum mechanics was the exciting, newly discovered fundamental theory of nature. He believed that one had to accept all its implications and fashion one's world-view in accordance with them. For Einstein, a non-deterministic theory, such as quantum mechanics, could not possibly be a fundamental theory of nature. Quantum mechanics had to be accepted, because it worked, but this acceptance was grudging and tentative. One had to keep looking for the really fundamental theory of nature, a theory that would show quantum mechanics to be a limiting case, a statistical approximation.

Realism and strict causality are distinct ideas. One can be a realist and still allow for a measure of randomness in the determination of events. But Einstein refused to take this route. This stubborn attachment to strict causality puzzled many of his fellow physicists. In other arenas Einstein showed an uncanny capacity to give up cherished notions. His own discoveries introduced major paradigm shifts. Furthermore, the statement "God does not play dice" is problematic: The doctrine of strict causality implies a God that does not allow for randomness, and in doing so it also implies that after the initial act of creation God does absolutely nothing in relation to the world—hardly an attractive concept of the Divine!

There are weighty arguments in favor of realism. It appeals to common sense, and, in the context of science, it works, at least up to a point. But the assumption of strict causality can be viewed as arbitrary. Why was Einstein so attached to it? Why did he steadfastly refuse to give it up in spite of the presence of randomness as an essential feature of quantum mechanics, a theory whose spectacular success he fully realized?

The present chapter expresses my own thoughts about this issue. To begin, strict causality was just one feature of Einstein's world-view, a view that was largely based on the findings of Special Relativity. As we will see, these findings and strict causality are interrelated. To see the interrelationships, we proceed to discuss Einstein's Special Theory of Relativity. We will come back to the issue of strict causality in the last section of this chapter.

2. What Is Real?

The answer to the question "What is real?" reflects the world-view of a civilization, a culture, or a system of thought. In the dominant philosophical systems of the ancient Greeks, as well as in most religions, the answer to the question "What is real?" is given in terms of "levels of being." Some elements are more real than others, and the relationships among the different elements form a distinct hierarchy.

For Plato "the Good" is the supremely real. The Forms, which occupy a lower level than the Good, are eminently real, and the sensible world, the objects in space and time reported to us through the senses, have only a shadowy kind of reality. This idea was refined by the neoplatonists in the second and third century A.D. For Plotinus, the greatest of the neoplatonists, the sequence begins with "the One" and continues through the Nous (the Universal Mind or Intelligence), the World Soul, our individual souls, and nature and ends with the sensible world. The sensible world is the least real of all—its reality is a borrowed one.

In Western religious systems God is at the top of the ladder. Below Him are high beings, such as archangels and angels; the hierarchy continues with humanity, then animals, plants, and finally inanimate matter. The idea of hierarchy applies to individuals as well: In both Platonic and religious thinking the soul is higher, more real than the body.

But Greek philosophy contains another strand of thought, materialism. Materialism can be traced back to Leucippus and Democritus in the fifth century B.C. It was passionately promoted by Epicurus in the second century B.C., and eloquently expressed by the Roman poet Lucretius around 55 B.C. Lucretius stated the essence of the materialistic doctrine as follows:

> All nature as it is in itself consists of two things—bodies and the vacant space in which the bodies are situated and through which they move in different directions . . . nothing exists that is distinct both from body and from vacuity.[2]

Throughout the Middle Ages this doctrine was considered heresy par excellence, and Epicurus was assigned a place of honor in Dante's hell. With the rise of modern science in the seventeenth century, however, materialism was ready for readoption.

3. The Mechanistic Universe

> It seems probable to me that God in the Beginning formed Matter in solid, massy, hard, impenetrable, movable Particles, of such Sizes and Figures, and with such other Properties, and in such Proportion to space, as most conduced to the end for which he formed them; and as these primitive Particles being Solids, are incomparably harder than any porous Bodies compounded of them; even so very hard as never to wear or break in pieces;

no ordinary Power being able to divide what God himself made one in the first Creation.[3]

This statement of Newton's reveals a concept of reality that contains two levels, God and nature. Since God created nature, His reality is supreme, certainly higher than the reality of the nature He created. But, as Pierre Simon de Laplace and others noted in the following centuries, the concept of "God" was logically extraneous to the Newtonian paradigm. The paradigm did not need a God to create or sustain it. Though Newton himself was deeply religious and devoted more time and energy to the study of theology than he did to the study of physics, the system he introduced was the foundation of a world-view that was not only materialistic but mechanistic: nature as a perfect clockwork, a reality which contains just the clockwork itself, without a designer or a craftsman who built it.

Once the unnecessary hypothesis of God's existence was dispensed with, the stage was set for conceiving of the universe as a mere collection of particles moving in space. The immutability and supremacy of God were replaced by the immutability and supremacy of the laws of nature—Newton's laws. In the physics of the eighteenth and nineteenth centuries the clockwork-universe paradigm reigned unchallenged. Physicists, as individuals, may or may not have believed in God, but the working paradigm of physics was the atheistic, materialistic one.

The materialistic concept of nature was modified during the second half of the nineteenth century, when it had become clear that the austere Newtonian universe, composed only of particles and void, would not do. This modification did not change its basic mechanistic character. It merely added to nature a new type of ingredient, called a "field." The need for this addition arose as follows:

The Newtonian framework included the problematic notion of "action at a distance." For example: The earth is here, and the moon is there. How does the earth exert a gravitational pull on the moon? Well, it just does, acting at a distance. But those who found the notion of "action at a distance" disturbing proposed another explanation: Assume that the earth, merely through its massive presence, generates a "gravitational field" in the space around it. This field mediates the gravitational relationship between the earth and the moon. The moon is pulled toward the earth not because it feels the faraway presence of the earth but because it feels the immediate presence of the earth's gravitational field right where it is.

From the point of view of eighteenth-century physics the concept of the gravitational field is not essential. If one accepts the awkward notion of action at a distance, one need not invoke the concept of a gravitational field. When we consider the complex arena of electromagnetic phenomena, however, the concepts of "electric fields" and "magnetic fields" prove indispensable. In fact, the crowning achievement of nineteenth-century physics is a set of equations, formulated by James Clerk Maxwell, which delineates the intricate relationships between these two types of fields, as well as their relationships with the electric charges and currents that give rise to them. Most of the technological inventions of the past century, from electric generators and radio to television and computers, are based on these equations.

What is an electric field? Fundamentally, it is a potentiality to influence the motion of electrically charged particles. If an electric field is present in a certain region in space, and a charged particle happens to be there, it will feel a force. And if no particle is present at that region? Well, in that case nothing happens. That is why the field itself is essentially a potentiality.

This brief account of the addition of "fields" to the mechanistic universe is conceptual rather than historical; it is written with the benefit of hindsight. The historical development was not as simple as the above paragraphs indicate. In the seventeenth and eighteenth centuries there were numerous attempts to avoid the notion of action at a distance by explaining the force of gravity as a transmission of forces through contacts among the particles of a hypothetical omnipresent subtle form of matter, called "ether." In the nineteenth century the ether was invoked once again in order to avoid the introduction of "fields" as a new type of entity by explaining the effects of electric and magnetic fields in terms of material particles. Numerous models were constructed to explain these effects as results of stresses and strains in the ether. These models ran into difficulties; the properties that the ether had to be endowed with to account for the observed phenomena were hard to believe. All of these models were discarded once Special Relativity appeared on the scene, because Special Relativity turned out to be inconsistent with the assumption of the existence of an omnipresent ether.

Toward the end of the nineteenth century the mechanistic paradigm of classical physics reached its high point: The real entities in the universe are particles and fields, existing and changing according to Newton's laws of mechanics and Maxwell's laws of electromagnetism. To explain a phenomenon such as free fall, the tides, or the magnetism of the earth, we have to show how it arises as a configuration of particles and fields subject to these laws. In principle all phenomena are explainable within this framework; hence chemistry is reducible to physics, biology to chemistry, and psychology to biology. And consciousness, which is hardly significant in the immense scheme of the largely inanimate universe, is but one specific item of study within psychology.

4. What Is Time?

The mechanistic paradigm of the late nineteenth century was the one Einstein came to know when he studied physics. Most physicists believed that it represented an eternal truth. But Einstein was open to fresh ideas. Inspired by Mach's critical mind, he demolished the Newtonian concepts of space and time and replaced them with new, "relativistic" concepts.

The full extent of Einstein's thought and achievements lies outside of the scope of this book. We will explore in detail, however, his first major breakthrough, his discovery of the *relativity of simultaneity*. This discovery revolutionized the understanding of space and time, put the question "What is real?" in a new light, and had a bearing on the issue of whether or not "God plays dice."

The question "What is real?" is intimately related to the question "What is time?" This is especially true in the currently held paradigm of Western civilization: For Plato "the Real" was the timeless Forms, or Ideas, such as Beauty, Love, or the Number Three. For us, however, "the real" is transient things and events, such as tables and chairs, lightning bolts, or our bodies. Hence our concept of "the real" is intertwined with our concept of time.

Well, then, what is time? St. Augustine begins his investigation into the nature of time with a perceptive comment: "I know well enough what it is, provided that nobody asks me; but if I am asked what it is and try to explain, I am baffled." He goes on to argue and question as follows: The past does not exist, because it is already past; the future does not exist, because it hasn't happened yet. Only the present is real. The present, however, has no duration; it is merely the demarcation line between past and future. And yet we do have an awareness of periods of time: We have an awareness of something taking a long time, and something else taking only a short time. How is such awareness possible? If that which exists, namely, the present, has no duration, how can we be aware of "a long time"? How can we be aware of something that does not exist? Augustine's response to the question is an insight into the nature of time. As we experience "a long time," he writes, "It is not future time that is long, but a long future is a long expectation of the future; the past time is not long, but a long past is a long remembrance of the past." St. Augustine concludes: "It is in my own mind, then, that I measure time. I must not allow my mind to insist that time is something objective."[4]

St. Augustine's lucid analysis is a forerunner of the deep, if obscure, analysis of Kant, an analysis that led to the conclusion that time is a category of the human mind. The paradigm we currently live by regarding the relationship between time and "the real" is still informed by St. Augustine's insight: Neither the past nor the future is real; only the present is. But there is a crucial difference between our view and St. Augustine's. Augustine's basic premise is that the only reality is God. This basic premise is not shaken by the discovery that time is unreal. By contrast, God is not a part of the scientific world-view. And so, if the past and the future do not exist, we are left to conclude that *"the real world" is the world as it is at the present moment.* It follows that *the real world has no duration.*

Furthermore, if this is indeed the case, *we are never aware of the real world.* We can't possibly be aware of the real world because it always takes time, however short, for the carriers of sense impressions (light, sound) to reach our bodies. Even in the case of immediate perceptions, such as touch or taste, it takes time for the nervous system to process the sensory input and create an experience. In our awareness of the physical world we are always behind. By the time we experience it, we experience it not as it is but as it was. Nevertheless we are clearly convinced (and this conviction is part of our world-view) that in spite of this gap between existence and awareness, there is a real world out there, the world of the present moment. It consists of the collection of all the events that take place right now. Thus *the real world is a collection of simultaneous events.*

5. Special Relativity: The Basic Postulates

The conclusion of the last section will help us appreciate the significance of the paradigm shift introduced in 1905 by Einstein's Special Theory of Relativity. The theory is based on two postulates. The first, called *the principle of relativity*, dates back to Galileo and Descartes.

Imagine that you are traveling in a sealed train car; you cannot look outside. Imagine, further, that the ride is so smooth that you don't feel any vibrations. When you entered this sealed car, the train was in motion, so you assume that it still is; but, after a while, you may begin to wonder: Is it still moving? This may lead you to ponder: Is there a way to determine whether the train is moving or not without leaving the car or looking out?

Well, it depends. If the car is accelerating, you will feel pushed back, and you will know that it is moving. If the train moves around a bend, you will feel pushed sideways, and you may even lose your balance—no doubt, the train is indeed moving. If, however, the train keeps moving at a constant speed and in a straight line, there is no way to tell. *But if there is no way to tell, even in principle, then the question "Am I moving or am I at rest?" is meaningless.* When we think about trains moving on the surface of the earth, the overwhelming presence of the earth makes it natural to assume that the earth is "really" at rest. But this is a kind of optical illusion. As far as the laws of nature go, they are the same on the platform and in the steadily moving train. And the motion is really relative. I, standing on the platform, consider myself at rest, and I see you moving at a constant velocity. But you, standing inside the train car, have every right to consider yourself at rest and think of me as moving at a constant velocity in the opposite direction. Who is "really" at rest? There is no such thing as being "really" at rest, or "really" in motion. Relative to me, you are moving. Relative to you, I am in motion. For inertial observers, i.e., observers whose velocities stay constant, that's the end of the story.

While this first postulate, the principle of relativity, makes sense, the second postulate does not. The second postulate, *the principle of the constancy of the speed of light*, states that the speed of light is the same regardless of the speed of the source which emits the light and regardless of the speed of the observer who measures it.

The value of this constant speed of light is about 186,000 miles per second. Light is really moving fast! The remarkable thing, however, is not the value but the constancy of this value when measured under different conditions. Common sense tells us that this cannot be true: Suppose I travel on a train that moves at 100 miles per hour and throw a ball toward the front of the train at 30 miles per hour. Suppose, furthermore, that the train is passing through a station without stopping or slowing down. If the stationmaster, standing on the platform of his station, measures the speed of the ball relative to him, the result of his measurement would be the sum of the speed of the train and the speed of the ball relative to the train, i.e., 130 miles per hour. The principle of the constancy of the speed of light says, however, that if I "throw" a pulse of light instead of a ball, this procedure of adding up the velocity of light and the velocity of the train yields a wrong result; the measurement of the speed of light by the stationmaster would yield the same result as the measurement of the passenger on the train—as if the train were not moving at all!

"Throwing" pulses of light could mean something as simple as holding a flashlight and flicking it on and off. The Michelson-Morley experiment of 1886, which surprised the world of physics by demonstrating this constancy of the speed of light, was much more refined, but its idea and its results are correctly illustrated by our simple example.

6. The Relativity of Simultaneity

Einstein came to accept this second postulate not because of the Michelson-Morley experiment but through an analysis of Maxwell's theory of electromagnetism. The speed of light appears as a constant in Maxwell's equations; and if, following the principle of relativity, one assumes that Maxwell's equations are equally valid in all frames of reference that are moving at a constant velocity, then this constant, the speed of light, must, indeed, be constant, i.e., have the same value in all such frames. Following this line of thinking, Einstein accepted the second postulate in spite of its apparent absurdity. This acceptance was an ingenious leap of faith. It allowed Einstein to take the position that both postulates, the principle of relativity and the constancy of the speed of light, are true, and then show that if we are willing to radically change the way we think about space and time, both postulates can indeed be a part of a consistent world-view. Let us assume the validity of the two postulates, he argued, and see what follows.

The first concept that came under Einstein's scrutiny was the concept of simultaneity. How do we determine whether or not two events are simultaneous? This is easy if both events occur right here: If I saw them happening at the same time, then they did happen at the same time. But what about two events that are taking place far away from each other? What if one of them took place right here and the other on the moon? In this case, if I saw them at the same time, then, certainly, they did not occur at the same time: The event on the moon took place a second and a quarter before the event here, because it takes light that much time to travel from the moon to the earth. The definition of simultaneity must take into account the time gap between the happening of an event and the arrival of light from the event to the observer. To take this time gap into account, let us say, with Einstein, that events are simultaneous if an observer *situated halfway between their locations* sees them at the same instant.

Using such "midway observers" we can synchronize two clocks situated apart from each other. The clocks need to be set so that an observer situated midway between them will see them showing the same time. If these clocks are of identical construction, and neither breaks down, we can assume that, once set, they will continue to be synchronous as time goes on. This is all rather straightforward. Our last assumption holds true if neither clock is under the influence of gravity, or, in the language of the *General* Theory of Relativity, if both clocks are in regions of spacetime that are flat. We will limit our discussion to this case.

We are now ready to begin the process that will lead to "the relativity of simultaneity." The process involves one of Einstein's favorite intellectual journeys, a thought experiment. A thought experiment is an experiment that can be performed in principle (i.e., its performance is allowed by the laws of nature) and yet, because of the limitations of our technology, cannot be performed in practice. However, even though we cannot do the experiment, nothing prevents us from *thinking* about it, and such thought can lead to far-reaching conclusions, as we will see. We will describe our thought experiment, which is similar to the one proposed by Einstein, in the form of a story.

Once upon a time two astronauts, Peter and Julie, were traveling in their respective spaceships way out in space, far away from stars and planets. Both spaceships had clocks at each end; each pair of clocks was synchronized: Peter and Julie, standing in the middle of their vehicles, acted as "midway observers."

The two spaceships approached each other. They traveled in precisely opposite directions, and, because of Peter's carelessness, they almost collided head-on. Fortunately, the spaceships just missed each other as they zoomed past. The encounter, while not disastrous, was not uneventful. This is how Julie told me about it (see Figure 2.1):

"I was just standing there, in the middle of my spaceship, which was at rest, when Peter's spaceship appeared, zooming toward me at an enormous speed. I was relieved when I realized that it did not hit my spaceship, but right at the instant his spaceship was side by side with mine, something strange happened: Two pieces of space debris, two small rocks, hit our vehicles. It happened at the

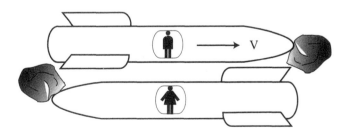

Figure 2.1   The situation at 12:00 according to Julie. The two rocks are hitting the ends of both spaceships, as Peter, standing in the middle of his spaceship, zooms past Julie, who is standing in the middle of hers.

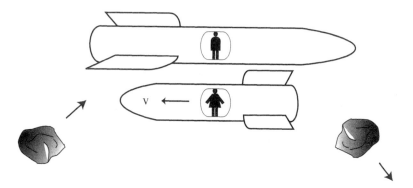

Figure 2.2   The situation at 12:00 according to Peter. One rock has already hit the front of Peter's spaceship and the rear of Julie's, while the other is about to hit the rear of Peter's spaceship and the front of Julie's, as Julie, standing in the middle of her spaceship, zooms past Peter, who is standing in the middle of his.

exact moment our ships were perfectly parallel, the back of his ship almost touching the nose of mine and the nose of my ship almost touching the back of his. It was at this precise moment that the fronts and rears of our two ships were hit simultaneously by the two rocks. The rocks weren't big enough to penetrate our hulls, but they jolted our ships pretty violently—violently enough to stop both of my clocks. Later Peter told me that the rocks hit hard enough to stop his clocks too. The strangest part of it is that the jolts occurred at the exact same time!"

"And how do you know that the two jolts happened at the same time?" I asked.

"Well," Julie responded, "as I just said, when my two clocks stopped, they showed twelve o'clock. As a matter of fact, since they're still broken, they still do."

"That's strange," I responded. "I talked to Peter a little while ago, and he told me that the two jolts were not simultaneous. One of them took place a little before noon and the other a little after noon." (See Figure 2.2)

Julie pondered this piece of information for a long time. Finally she said: "What about the broken clocks at the front and rear of his vehicle? They stopped when the rocks hit. What do they read?"

When I answered, she could hardly believe her ears: "*The clock in the front of his spaceship shows a fraction of a second before 12:00; the one at the rear, a fraction of a second after. According to his clocks, at 12:00 the jolt to the front of his spaceship already happened, and the other jolt hadn't happened yet.*"

It was not only Julie who could not believe her ears. It is, indeed, hard to accept the relativity of simultaneity. Remember, "the world now" is what we consider real. According to Special Relativity, however, the collection of events that comprises "the world now" is different for different observers. Julie's "real world at 12:00" includes the two rocks in the acts of hitting the spaceships, while Peter's "real world at 12:00" includes neither of these events. In his "world at 12:00" event one jolt already happened, and the other is still in the future. What is *the really real world at 12:00* like? Einstein's message is that the concept of a really real world at 12:00 is meaningless. As we show in Section 1 of Appendix 1, the relativity of simultaneity is an unavoidable conclusion of the two basic postulates of Special Relativity.

In our thought experiment the differences between Peter's and Julie's clock readings for events A and B amounted to very small fractions of a second. This is due to the fact that the two events occurred in close proximity to each other. The relativity of simultaneity can lead to very large differences in clock readings if the two events are far apart. Let us assume, for example, that the encounter between Peter and Julie took place very far from the solar system, light years away from earth. Meeting and reminiscing years after the encounter, they could have compared the time of the encounter with the times of events on earth. Peter could have said to Julie, "Remember that near collision we had? It happened on December 30, 1989, the same day my poor uncle died." "No, no," Julie would have responded, "it happened on March 4, 1990, when my nephew was born!"

Let us conclude this section with the following observation: If you compare Figures 2.1 and 2.2, you may notice something strange about the lengths of the two spaceships. In Figure 2.1 they are of equal length, while in Figure 2.2 Peter's spaceship is longer than Julie's. This is not an error. It follows from the two basic postulates of Special Relativity that length is relative! If you measure the length of a spaceship (or any other object) while it is moving relative to you, the result is less than the length of the same object measured when it is at rest relative to you. This is called "length contraction." The explanation of how length contraction comes about and how it is related to the relativity of simultaneity is given in Section 2 of Appendix 1.

7. An Ultimate Speed

The relativity of simultaneity is fascinating. But what does it have to do with Einstein's adherence to strict causality?

Before we respond to this question, let us introduce one more consequence of the two basic postulates of Special Relativity: The speed of light is the greatest speed that occurs in nature. Nothing can move faster than light. Light itself, and other forms of radiation, can't help but move precisely at the speed of light; this is what the second postulate tells us. By contrast, material particles and objects, such as an electron, an atom, a spaceship, or a human body, can approach the speed of light but can never reach it. And since everything in nature is either matter or radiation, nothing can move faster than light. Furthermore, since information is transmitted from one place to another by some agent, the speed of light is the ultimate speed for the propagation of signals.

The speed of light is thus the ultimate speed by which a cause can bring about an effect. If one event A is the cause of another event B, there must be enough of a time separation between them to allow a signal traveling at the speed of light from A to reach B. This amounts to a kind of separability: Events that are separated in space can be related as cause and effect only if there is enough of a time separation between them as well. This characteristic of nature is called "Einstein separability" or "locality."

8. The Eternal Present

The relativity of simultaneity is, in Alfred North Whitehead's words, "a heavy blow to the classical scientific materialism, which proposes a definite present instant at which all matter is simultaneously real."[5] What Whitehead calls "the classical scientific materialism" is the world-view we live by. The negation of our intuitive concept of the real by Special Relativity demonstrates that whatever "the real" is, it is not what we think. If "the world now" cannot be "the real world," what is? This was Einstein's dilemma.

St. Augustine was led into his investigations of the nature of time through the bothersome question "What did God do before He made heaven and earth?" "My answer," he wrote, "is not 'He was preparing Hell for people who pry into mysteries.' This frivolous retort has been made before now, so we are told, in order to evade the point of the question."[6] Contemplating the difference between time and eternity, he concludes that before Creation *there was no time.* Addressing God, he expresses his conclusion as follows:

It is therefore true to say that when you had not made anything, there was no time, because time itself was of your making. And no time is co-eternal with you, because you never change; whereas, if time never changed, it would not be time.[7]

The human perspective of dividing events into past, present, and future events is irrelevant from the perspective of eternity, which is God's perspective:

It is in eternity, which is supreme over time because it is a never-ending present, that you are at once before all past time and after all future time. ... Your years neither go nor come, but our years pass and others come after them, so that they all may come in their turn. Your years are completely present all at once, because they are at a permanent standstill.[8]

The seeming unreality of past and future is not really problematic for St. Augustine. The division into past, present, and future is a creation of the human mind. From God's perspective there is no such division. Hence all events, past, present, and future, have equal status as far as their reality goes. They may or may not be "really real," but the events of the present moment are not more or less real than the events of the past or the events of the future.

A similar conclusion regarding the reality of past, present, and future events follows from the relativity of simultaneity. If events that are present for one observer are past or future for another, the question of whether or not they are "real" will not be settled by checking on whether they are "now events" or not. The appeal to the perspective of God is not part of the current scientific paradigm, but Augustine's point of view regarding the reality or unreality of physical events is forced upon us by the relativity of simultaneity. Einstein resolved his dilemma by concluding that fundamentally all events, past, future, and present, are on equal footing, that they are all real. Time is an illusion.

A month before he died, Einstein received the news of the death of his life-long friend Michele Besso. His letter of condolence to the Besso family contains the following remarkable statement:

Now he has departed from this strange world a little ahead of me. That signifies nothing. For us believing physicists the distinction between past, present, and future is only a stubbornly persistent illusion.[9]

9. The Case for Strict Causality

As you may recall from the beginning of this chapter, this long excursion into the Special Theory of Relativity was motivated by the wish to understand Einstein's commitment to the tenet of strict causality.[10] Now we are in a position to appreciate the source of this commitment.

Following the indication of Special Relativity, Einstein ascribes reality to all the events in spacetime. All events, regardless of whether they are past, present, or future events from the point of view of particular observers, are on the same footing; they are all real. The apparent unreality of the past and the future is merely the result of the way our minds process impressions.

Now, if we are sure of anything, it is that *the past is fully determined.* Whatever happened, happened—there is no randomness about that. But if all events are to be treated on the same footing, we must extend this ontological certainty to the present and future as well. If time is an illusion, a mere human perspective, the perspective of a limited mentality facing the vastness of the "eternal present," then, in a sense, *the present and the future have already happened, just like the past.*

The belief in strict determinism is, then, one element in a consistent worldview that is based, on the one hand, on the belief in an independently existing reality and, on the other, on the findings of Special Relativity. As we will see, however, according to the principles of quantum mechanics the physical universe is not deterministic. This is one reason for Einstein's refusal to accept quantum mechanics as a fundamental theory of nature.

There is, however, another reason. Einstein's belief in realism included the commonsense assumption that the description of a physical system requires a single model. This belief applies to all systems, large and small, even to systems that belong to the quantum domain. However, when Niels Bohr struggled to understand the quantum theory he discovered that this is not the case. Two years after Heisenberg's lecture at the University of Berlin he introduced his "framework of complementarity," according to which a full description of a quantum system involves, in general, two seemingly contradictory conceptual models. Let us turn now to the exposition of the great challenge to Einstein's paradigm: Bohr's world-view and its mainstay, the framework of complementarity.

3. The Call of Complementarity

Bohr and Einstein were both puzzled by "the wave-particle duality," the fact that light, as well as electrons, behaves sometimes like waves and sometimes like particles. Unlike Einstein, however, Bohr approached the problem without a priori metaphysical assumptions. He solved the puzzle by introducing his "framework of complementarity," which is exemplified by Heisenberg's "uncertainty principle."

~

That which is in opposition is in concert, and from things that differ comes the most beautiful harmony.

—Heraclitus

1. Bohr Introduces the Framework of Complementarity

Niels Bohr's first major presentation of his framework of complementarity took place on September 16, 1927, at an International Congress of Physics at Como, Italy, a congress from which Einstein was absent. The discovery of this framework was the successful culmination of fourteen years of intense and frustrating intellectual struggle, and Bohr's Como lecture is considered a landmark event in the history of physics.

Like the crowds that listened to Abraham Lincoln's Gettysburg address, however, Bohr's audience was slow to realize the significance of what they had heard. Oskar Klein, Bohr's assistant at the time, remarked that the lecture was "so short nobody could understand it really."[1] Moreover, Bohr was notoriously obsessive about presenting his ideas with the utmost precision of expression; as a result his formulations were often obscure. And there might have been also a more mundane reason for the cool reception. I had the opportunity to hear Bohr lecture when he visited the Weizmann Institute of Science in Rehovoth, Israel, in May 1958. The largest auditorium was filled to capacity with people thrilled at the opportunity to hear Bohr speak. We listened in awe and in complete silence to a long lecture but couldn't understand more than a few words because of Bohr's heavy Danish accent.

Bohr's idea of complementarity is subtle and hard to grasp. Referring to "Bohr's principle of complementarity," Einstein wrote in 1949: "The sharp formulation ... I have been unable to achieve despite much effort that I have expended on it."[2]

25

In this chapter we may not "achieve the sharp formulation" either, but we will clarify the thrust of the idea. Our first task is to understand the difficulty that, with his idea of complementarity, Bohr claimed to have resolved.

2. Light: Particles or Waves?

What is the fundamental nature of light? Newton, who believed that the physical universe consists of particles moving in empty space, hypothesized that light is made of very small particles. When these particles impinge upon the retina of the eye, we see. The perception of different colors corresponds, in Newton's view, to differences in the size of the particles. For example, particles corresponding to perception of blue are larger than those that produce perception of red.

This theory explained all of the properties of light that were known in Newton's time. For example, light that hits a mirror is reflected in such a way that the angle of reflection is equal to the angle of incidence, just like a ball bouncing off the floor (see Figure 3.1). Other properties of light can be similarly explained by analyzing the behavior of little particles in analogous situations.

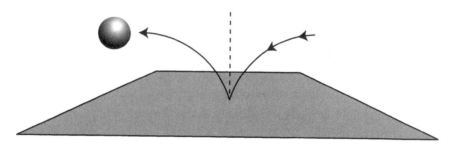

Figure 3.1   For light, just as for a bouncing ball, the angle of reflection is equal to the angle of incidence.

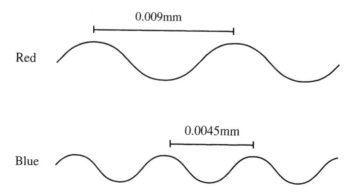

Figure 3.2   The "wavelength" of a wave is the size of the repeated pattern. According to the wave model of light, different colors correspond to waves of different wavelength.

Incident waves

Reflected waves

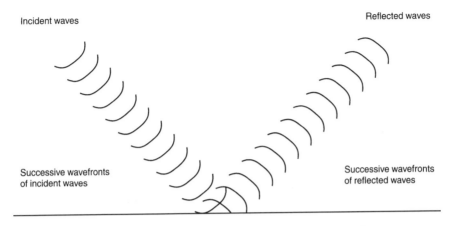

Successive wavefronts
of incident waves

Successive wavefronts
of reflected waves

Figure 3.3   For waves, just as for particles, the angle of reflection is equal to the angle of incidence.

The success of the particle model led Newton and his followers to declare that light is made of particles, but the acceptance of the particle model was not universal. Christian Huygens, a contemporary of Newton, suggested an entirely different model, in which light is described as a wave phenomenon. Rays of light, moving in space, are, according to Huygens, tiny waves propagating in an omnipresent medium called "ether." Different colors correspond to waves of different "wavelengths" (see Figure 3.2). Odd as it may seem, it turned out that this wave model was just as successful in explaining the properties of light which were known in the seventeenth century. The law of reflection, for example, is just as true for waves as it is true for particles (see Figure 3.3). Well, then, if both of these models account for its characteristics, what is light *really* like?

For over a hundred years there was no definitive answer. Nobody doubted that one of the models was really right, and, essentially because of his prestige, most physicists sided with Newton. A seemingly definitive answer was discovered at the beginning of the nineteenth century, when Thomas Young performed an experiment whose results could only be explained on the basis of the wave model and not the particle model.

In Young's experiment (Figure 3.4), a beam of monochromatic light is directed toward a diaphragm into which two narrow slits have been cut close together. The light passes through the slits and proceeds toward a screen. The curve drawn to the right of the screen in Figure 3.4 indicates the brightness of light on the screen. An actual photograph of what such a screen looks like is shown in Figure 3.5. We see a succession of bright and dark bands. How can this pattern be explained?

The dark and light bands are the result of a property of waves known as *the principle of superposition*: If two waves pass through the same place, the resulting vibration of the particles that occupy this place will be equal to the combined effect of the two waves; at each instant the displacements due to the two waves are added up, or "superposed" upon each other.

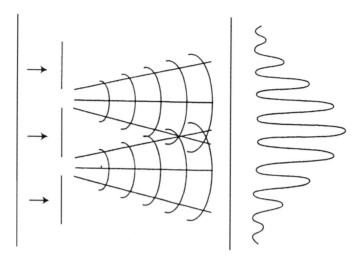

Figure 3.4   Young's experiment: A beam of monochromatic light hits a diaphragm with two narrow slits, close together. The light that passed through the slits proceeds toward a screen. The curve drawn to the right of the screen indicates the brightness of light on the screen.

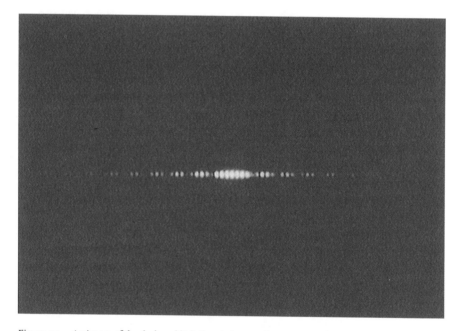

Figure 3.5   A picture of the dark and light bands in a modern version of Young's double-slit interference experiment. In this current version of the experiment the light is emitted by a laser. The photograph is reproduced here courtesy of Roger Williams.

Yesterday was a breezy fall day, near the peak of foliage season in upstate New York. In the late afternoon I took a walk by Taylor Lake, on the campus of Colgate University. The lake is famous for its swans, but I was paying attention to the leaves that floated gracefully on the surface of the water. There were small waves traveling along the surface, and, watching the leaves as the waves passed them, I noticed that some leaves were moving up and down, while others, close to them, were hardly moving at all! My curiosity was aroused; I watched two waves approaching a leaf from two different directions, so that at a certain instant it was under the influence of both. Each wave, acting separately, would have caused the leaf to be displaced upward by about an inch. Following the principle of superposition, the actual displacement of the leaf, due to the combined effects of both waves, was about two inches. The two waves "interfered constructively" with each other: They pushed the leaf in the same direction, and their effects added up. Then I watched another leaf, under the influence of two different waves. This time one of the waves would have imparted a displacement of an inch upward, while the other would have imparted a displacement of an inch downward. In this case, again following the principle of superposition, the effects of the two waves canceled out, and the leaf did not move at all. This is called "destructive interference." So, some of the leaves moved up and down under the influence of two waves that reinforced each other, and some leaves were under the influence of two waves that canceled each other out. This was why the waves moved some of the leaves and left other leaves at rest!

The principle of superposition explains the succession of dark and bright bands in Young's experiment. The brightest spots on the screen are at the locations where the two waves coming from the two slits interfere constructively, and the dark bands are at the places where the two waves interfere destructively. And, of course, between the brightest locations on the screen and those that are totally dark there are areas with different degree of grayness, corresponding to interference that is partially constructive and partially destructive.

The crucial point is this: The principle of superposition applies to waves but not to particles! The orderly appearance of bright and dark bands in Young's experiment cannot be explained by the particle model of light. More than a hundred years after his death, Huygens's wave model was proven right, and Newton's particle model had to be abandoned.

The evidence of Young's experiment was clinched in the 1860s, when James Clerk Maxwell discovered, through a brilliant mathematical analysis, the precise nature and structure of light waves. He demonstrated that they are a combination of electric and magnetic fields, coupled together and changing in unison in such a way that the wave moves, in vacuum, at the enormous speed of 186,000 miles per second.

3. Light: Particles *and* Waves?

Nothing in physics is definitive, however. As the nineteenth century drew to a close, Maxwell's electromagnetic wave theory of light encountered two stubborn problems. The first had to do with glowing objects. When an object, such as a piece of metal, is heated up, it starts glowing. i.e., emitting light. But Maxwell's theory cannot explain the observed characteristics of the glow: Calculations based on this theory lad to the conclusion that the amount of energy emitted should be infinite—an obvious absurdity!

The second stubborn problem was "the photoelectric effect": Under certain conditions a light shining on a piece of metal makes the metal emit electrons. This means that the energy of the light, impinging on the metal, gets concentrated on electrons and gives them mighty shoves. It is as if the energy of beach waves along a hundred miles of shore gets concentrated on a few big boulders, sending them hurling up into the air. But according to Maxwell's theory, light waves move continually, spread over space, and there is no reason for their energy to be concentrated like this. In other words, according to Maxwell's theory the photoelectric effect should not happen at all!

The dawn of a new century would also be the dawn of the quantum revolution. In 1900 Max Planck showed that the first problem would be resolved if light is assumed to be emitted from and absorbed by matter not continually, like waves, but discretely, in little packets of energy, which he called "quanta." Five years later Albert Einstein showed that the photoelectric effect can be accounted for if light is assumed to have this discrete nature not only when emitted or absorbed but also as it travels in space. These particle-like packets of energy were called "photons."

So, what is light made of, waves or particles? At the beginning of the twentieth century the situation seemed untenable: Certain properties of light could only be explained by assuming that it was a wave phenomenon, while other properties provided conclusive proof that it was made of particles!

This "wave-particle duality" has been a major puzzle since the inception of the quantum theory in 1900. The next step in the process that led to a solution was unexpected: In November 1924 a French prince, Louis de Broglie, presented his doctoral thesis in theoretical physics to the Faculty of Science at the University of Paris. The thesis contained a series of arguments in favor of the idea that not only light but also entities that were considered bona fide particles, such as electrons, have this puzzling dual nature, behaving in some respects like particles and in other respects like waves.

If de Broglie's conjecture is true, and electrons do behave at times like waves, can this wave aspect be observed experimentally? De Broglie suggested looking for wave characteristics in the passage of electrons through narrow openings, such as the spaces between layers of atoms in crystals. Such spaces can play, for electrons, the role that Thomas Young's slits play for light. The need for crystals, rather than slits, was due to the limitations of technology: In the 1920s it was impossible to create slits that were narrow enough, and close enough together, to demonstrate the wave nature of electrons. But de Broglie's idea of using crystals was a good one. An experiment that confirmed his hypothesis was conducted by two American physicists, Clinton Joseph Davisson and Lester Halbert Germer, two years later.

De Broglie's thesis of "matter waves" received a further boost in January 1927, when Erwin Schrödinger introduced a full-fledged mathematical theory called "wave mechanics" which accounted for the behavior of electrons in terms of waves. We will explore Schrödinger's theory in Chapter 4. Now, however, let us go back to Niels Bohr and his young friend and protégé, Werner Heisenberg, as the two of them are spending long hours in conversation, trying to decipher the contradictory pieces of evidence indicated by the expression "wave-particle duality."

4. What Can We Know About an Electron?

Heisenberg spent the summer and fall of 1926 in Bohr's institute in Copenhagen. Thinking hard, and talking for hours on end, Bohr and Heisenberg tried to understand the wave-particle duality as well as other disturbing aspects of the budding quantum theory. Their efforts continued into the winter of 1927. Many years later, Heisenberg recalled this period:

> Since our talks often continued till long after midnight and did not produce a satisfactory conclusion despite protracted efforts over several months, both of us became utterly exhausted and rather tense. Hence Bohr decided in February 1927 to go skiing in Norway, and I was quite glad to be left behind in Copenhagen, where I could think about these hopelessly complicated problems undisturbed.[3]

Left by himself, Heisenberg reviewed the situation anew and tried to discover what he and Bohr had missed. He was facing the following dilemma: On the one hand, looking at cloud chamber photographs such as Figure 1.1, it was obvious to him that electrons have trajectories. When he looked at the cloud chamber photographs he could actually see them. Having a well-defined trajectory is a property of a particle; waves tend to spread all over the place. On the other hand, quantum mechanics, a theory that was too successful in other respects to be discarded, did not allow for such trajectories. This was the impasse. After struggling with this paradox for days, he finally broke through:

> It must have been one evening after midnight when I suddenly remembered my conversation with Einstein and particularly his statement, "It is the theory which decides what we can observe." I was immediately convinced that the key to the gate that has been closed for so long must be sought right here.
> . . . We had always said so glibly that the path of the electron in the cloud chamber could be observed. But perhaps what we really observed was some-

thing much less. Perhaps we merely saw a series of discrete and ill-defined spots through which the electron had passed. In fact, all we see in a cloud chamber are individual droplets which must certainly be much larger than the electron. The right question should therefore be: Can quantum mechanics represent the fact that an electron finds itself approximately in a given place and that it moves approximately with a given velocity, and can we make these approximations so close that they do not cause experimental difficulties?[4]

Heisenberg realized that the quantum theory does not give us the right to assign simultaneous, *precise* values to both the position and to the velocity of electrons, but it does allow us to assign simultaneous, *approximate* values. The mathematical structure of the theory is such that one can find out a precise answer to either the question "Where is the electron?" *or* "How fast is the electron moving?" *but the theory does not allow us to have simultaneously precise answers to both questions.* If one wishes to have simultaneous answers to both questions, one has to settle for approximate answers; and a higher degree of precision in the answer to one question can only be reached at the expense of a lower degree of precision in the answer to the other. This is the essential idea of Heisenberg's celebrated uncertainly principle.

Look again at Figure 1.1. What do you see? When my friend Jane, the electrical engineer, looked at the photograph, she saw electron trajectories. But in the light of the principle that Heisenberg discovered, perhaps we can see something else: a sequence of approximate positions and approximate velocities. The fuzziness of the dots along the so-called trajectory is not only the result of the photographic process. Some of it is due to fundamental limitations on the possibility of observations. This has been recently confirmed, using advanced imaging techniques in which single atoms, rather than water droplets, were observed. It is, indeed, the theory that decides what we can observe. Armed with Heisenberg's uncertainty principle, we observe the same photograph differently.

Let us pause to introduce a new concept that applies to this principle. As it appears in everyday speech, the definition of the word "momentum" can be rather loose, but in physics it has a precise meaning: The momentum of a particle is its mass times its velocity. Intuitively, "momentum" represents the "amount of motion" of a particle. Because momentum is a product of two factors, mass and velocity, a massive object moving slowly can have the same "amount of motion" as a light object moving fast. And, of course, a massive object moving fast will have "a lot of momentum," while a light object moving slowly will have only "a little momentum."

Like velocity, momentum is a vector quantity: It has a direction as well as a magnitude. If, for example, a projectile is thrown upward (see Figure 3.6), its velocity points in the direction it was thrown, the direction of the bold arrow; the length of the arrow indicates, on an appropriate scale, the speed it was given. Momentum too can be represented by an arrow. The arrow points in the direction of motion, and its magnitude indicates the product of the mass and the speed. Since momentum is the product of mass and velocity, if we deal with a particle of known mass, such as an electron, knowledge of its velocity is equivalent to knowledge of its momentum.

As Figure 3.6 shows, a vector quantity, such as velocity or momentum, can be represented in terms of *components*. For example, the velocity of a projectile thrown upward at an angle has a vertical component and a horizontal component. Its motion can be thought of as a combination of these two components.

The concept of momentum is important because, unlike velocity, momentum is subject to a *conservation law:* If a system is left to itself, if no outside forces are acting on its particles, the total momentum of the system, i.e., the value of momentum we get by adding up the momenta of all its constituent particles, stays exactly the same, no matter what changes take place inside the system. This makes momentum measurements very useful. As we will see, the conservation of momentum enables us to find the momentum of an object indirectly, without measuring it. This will feature prominently in the Bohr-Einstein debate.

We are ready now to go back to young Heisenberg, huddled over his calculations at Bohr's institute in Copenhagen. We saw how, on that memorable night in February 1927, he discovered that there is, within the mathematical structure of the quantum theory, a kind of complementarity between the uncertainties in the simultaneously measured values of the position and velocity of a particle.

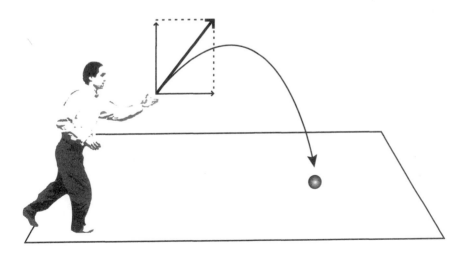

Figure 3.6   The bold arrow represents the velocity of a projectile at the moment it is released from the hand of the boy. The velocity is a vector, and it can be represented in terms of components. The vertical and horizontal components of the velocity vector are shown by the vertical and horizontal arrows. The horizontal arrow indicates the rate of the sideways motion, and the vertical arrow indicates the rate of the upward motion as the projectile moves at an angle.

Heisenberg formulated this complementarity relation in terms of momentum rather than velocity: Less uncertainty in the knowledge of position can only come at the cost of more uncertainty in the knowledge of the momentum, and vice versa. Heisenberg discovered not only the general idea but also the mathematical expression that relates the two uncertainties. This expression is the mathematical form of the uncertainty principle.

Having discovered the principle, Heisenberg became worried: What if it turns out that *experimentally* one can do better than the theory allows? What if there is a way to measure, simultaneously, both position and momentum to a degree of precision that the uncertainty principle says is impossible? He started reviewing standard experimental set-ups. One by one they proved, to his delight, to abide by his theoretical limit. Of course, he could not be absolutely sure that no one would ever find a way to beat his principle, but he found the results of his thought experiments encouraging. He was looking forward to sharing his discovery with Bohr upon Bohr's return from his skiing vacation in Norway.

5. Bohr Is Looking for the Right Words

The Archives for the History of Quantum Physics contain a series of interviews that Thomas Kuhn, author of the landmark book *The Structure of Scientific Revolutions*, conducted in the early sixties with some of the main players in the quantum drama that unfolded in the twenties. Talking to Kuhn in 1963, Heisenberg recalled Bohr's response to the discovery of the uncertainty principle: "Bohr, I would say, was a bit uneasy about the whole thing. He felt very strongly, 'Well, there must be something in it,' but at the same time he saw that one had not solved the problem of using the right words."[5]

"Using the right words"? What is *that* all about?

Bohr and Heisenberg agreed that the point of doing physics is to promote the understanding of the observed phenomena of nature. But what does "understanding" mean? Here Bohr was ahead of young Heisenberg. Heisenberg, still under the influence of Mach, mistook "understanding" for discovering a mathematical procedure that would successfully predict the outcome of experiments. For Bohr, however, "understanding" meant much more. It meant a *description* of the observed phenomena. Hence his search for "the right words."

Heisenberg's mistake was a fortunate one. It provided him with an appropriate frame of mind for making two momentous discoveries: the mathematical formalism of the quantum theory, and the uncertainty principle. For Bohr, however, this was not enough. He was looking for a conceptual framework, for "the right words." And he found them through his idea of complementarity.

Consider, with Bohr, an experimental arrangement for measuring the position of an electron along the vertical direction, i.e., measuring how high it is above a table. The arrangement consists simply of a narrow slit through which the electron passes. When it is in the slit, the uncertainty in its position is the size of the slit: We know that it is somewhere in the slit, and we are uncertain where. So, if a slit is very narrow, then the detection of an electron passing through it is a relatively accurate measurement of the vertical position, or height, of that electron. But what if we wish to measure accurately the vertical component of the *momentum* of the electron as it passes through the slit? (Since the momentum is a vector, it has both horizontal and vertical components.) If the partition with the slit is fixed to the table on which the experimental apparatus rests (set-up 1; see Figure 3.7), it cannot be done. But we can modify the experiment, hanging the partition on light, sensitive springs (set-up 2; see Figure 3.8). Now, as the electron passes through the slit, the vertical component of its momentum will cause it to bounce off an edge of the slit; as a result, the partition will start vibrating. By observing the vibration, we can deduce the magnitude of this vertical component. However, and this is the crucial point, *if the partition starts vibrating, the location of the slit relative to the table can no longer be determined.*

Bohr's conclusion was this: The experimental arrangement for measuring position (set-up 1) and the one for measuring momentum (set-up 2) are mutually exclusive. When we stretch our capacities for measurement to their limits, we end up either with an electron that has a well-defined position but not a well-defined momentum (set-up 1), or with an electron that has a well-defined momentum but not a well-defined position (set-up 2). Hence the properties of having a well-defined position and having a well-defined momentum are complementary and mutually exclusive. An electron can have either, but not both simultaneously.

Bohr's two set-ups illustrate Heisenberg's uncertainty principle and show its source: There is a profound difference between ordinary measurements and quantum measurements. In ordinary measurements, the measurement merely tells us the value of the measured quantity; it does not affect it. For example, when I measure the length of my desk and find it to be five feet, well, it is five feet, whether I measure it or not. In quantum measurements, however, the measuring apparatus has a profound effect on the state of the measured electron. If the electron is subjected to set-up 1, the set-up *imparts to it* a well-defined position and

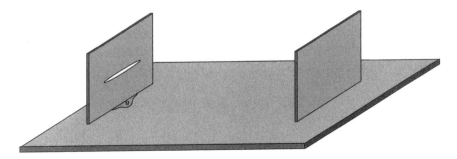

Figure 3.7   An experimental set-up for measuring the position of an electron as it passes through a slit (set-up 1).

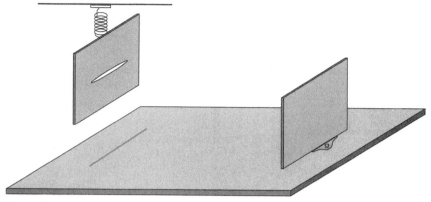

Figure 3.8   An experimental set-up for measuring the momentum of an electron as it passes through a slit (set-up 2).

ill-defined momentum; if it is subjected, instead, to set-up 2, the reverse is the case. And since an electron cannot be subjected simultaneously to both set-ups, the properties of having a well-defined position and having a well-defined momentum are complementary and mutually exclusive. The creative nature of quantum measurements will be explored in Chapter 4.

The uncertainty principle exemplifies one type of complementarity. A different type is exemplified by the relationship between *a quantum system in the process of being measured* and *a quantum system in isolation*.

To observe a quantum system, e.g., an electron, we need a measuring apparatus: An electron is too small to be observed directly, so we observe it indirectly, through its effect on the measuring apparatus. For example, in the case of set-up 2 we observe the vibrations of the partition. These were caused by the electron; hence the observation of the vibrations allows us to deduce its momentum.

During the process of measurement the electron and the measuring apparatus are interacting. While the interaction lasts, the two—the measured electron and the measuring apparatus—form one indivisible system, "a whole phenomenon." And, as we have seen, the results of the measurement depend not only on the state of the electron but also on the other part of "the whole phenomenon," the type of measuring apparatus that was chosen. For example, set-up 1 yields a well-defined position, while set-up 2 yields a well-defined momentum.

Once the measurement is over, the electron is isolated. The results of the measurement allow us to attribute a property to it—a well-defined position, if we used set-up 1, or a well-defined momentum, if we used set-up 2. At this stage, however, our description becomes abstract. It refers to an entity that is isolated, hence inaccessible. And precisely because it is isolated we can describe it in itself, separate from the rest of the universe.

We have, therefore, two mutually exclusive situations on our hands: Either we measure the electron, or we leave it alone. In the first case our description must include "the whole phenomenon": We cannot describe the electron itself, since there is no demarcation line between it and the measuring apparatus. In the second case we can describe the electron, but the description is abstract: The description cannot be verified, since it refers to an entity that is not monitored.

Bohr's idea of complementarity is complex, because it involves two types. One type is the complementarity that exists between different types of measurements, such as set-up 1 and set-up 2; the other type is the complementarity between the description of a system while it is being measured and the description of a system in isolation.

Through this interplay of complementarities, apparent contradictions disappear: The statements "the electron has a well-defined momentum" and "the electron does not have a well-defined momentum" are contradictory. But the statements "a measurement by set-up 2 yields a well-defined momentum" and "a measurement by set-up 1 does not yield a well-defined momentum" are complementary. And, once the measurement processes are over, this property of having or not having a well-defined momentum can be attributed, albeit abstractly, to the electron when it is on its own—an isolated system. Similarly, the statement "an electron is a wave phenomenon" and the statement "an electron is a particle" are contradictory. But according to the framework of complementarity, neither description is exhaustive; the very quest for a single description has to be given up. Two descriptions, one involving waves, the other involving particles, are necessary. Since they apply to different conditions, they are complementary rather than contradictory.

6. The Significance of the Framework of Complementarity

Thales of Miletus, who lived at the end of the seventh and the beginning of the sixth centuries B.C., is renowned and admired as the first scientist. Our knowledge about his teachings is confined to a few scattered comments by Aristotle, who attributed to him the statement that water is the first principle of nature.[6] This scanty record of Thales' scientific achievements can hardly serve as a basis for admiration. But Thales is worthy of our respect not because of his theory but because of what it implies. Thales was the first to theorize from the premise that *nature can be explained in rational terms* and that, furthermore, *nature can be modeled:* It is the achievable task of science to discover a conceptual model which corresponds to nature and accounts for the observed phenomena.

This premise has been the foundation of the scientific pursuit for over twenty-five centuries. From Anaximander and Aristotle to Einstein and Schrödinger, the great scientists of all ages have based their work on it. Bohr's framework of complementarity, however, is an explicit rejection of Thales' implicit claim. According to Bohr, *the entities that belong to the atomic and subatomic domains cannot be described by a single model.* This assertion is revolutionary enough, but Bohr couples this idea with another: According to him, not only are two complementary models needed, but what these models describe is not quite the quantum entities themselves.

Bohr and Einstein agreed that the task of science is to describe nature, but Bohr's understanding of what the expression "a description of nature" meant was different from Einstein's. Ordinarily, when we speak about describing nature, we assume that nature is there to be described. For Bohr, however, *"the description of nature" is the description of the collection of the objective poles of our experiences.* What does this strange expression mean?

Consider an example: I am looking at a nail as I hammer it into a piece of wood. The nail is at the focus of my awareness; it is the only thing I am focally aware of. In particular, I am not aware of myself, the one who sees the nail, although, obviously, I, the one who experiences, must be there for the experience of seeing the nail and aiming the hammer to take place. Every ordinary experience has this "subject/object structure." The objective pole of the experience is that which is experienced. The subjective pole is that which makes the experience possible, and it includes the experiencer. Now, a moment's reflection reveals that the demarcation line between subject and object can be fluid. As I keep hitting the nail, I can change the focus of my awareness and include both the hammer and the nail; if I so choose, I can even look at myself hammering, thereby including myself (or, rather, one of my selves—more on this later) in the objective pole.

Evidently, this demarcation line can even cut across the distinction between the animate and the inanimate. Just as the objective pole may include myself, as seen from the perspective of a deeper self, the subjective pole may include objects, such as a hammer. A blind person's walking stick belongs to his or her subjective pole almost all the time.

In the case of quantum measurements one may consider the measuring apparatus to belong to the experimenter's subjective pole, and the measured system to the objective pole; here, too, the demarcation line between the measured system and the measuring apparatus can change. Consider the case of a photon hitting an electron and then impinging onto a photographic plate. It is possible to consider just the electron as that which is measured; it is possible to consider the electron and the photon as the measured system; one may include the atom of the photographic plate which has been hit by the photon as part of the objective pole, and so on. According to Bohr, a *"description of nature" should be an account of anything that may happen in any of these divisions of perceptions and measurements into objective and subjective poles.*

But how, you may ask, is nature in itself, apart from these divisions? Isn't science supposed to account for that? No, says Bohr. The very question of whether it is meaningful to even speak about nature in itself is outside the domain of science. Bohr's student A. Peterson quotes him as saying: "It is wrong to think that the task of physics is to find out how nature is. Physics concerns what we can say about nature."[7] One profound difference between Einstein's understanding of the phrase "the description of nature" and Bohr's is the difference between the *ontic* and the *epistemic* views. "Ontic" means "relating to being"; "epistemic" means "relating to knowledge." According to Einstein's ontic view, the task of science is to describe nature as it is. According to Bohr's epistemic view, the task of science is to describe what we can know about nature, that is, the results of all conceivable perceptions and experiments.

Consider, as an example, an application of Heisenberg's uncertainty principle to the state of an electron. An ontic interpretation says, "An electron does not have, simultaneously, a precise position and a precise momentum." An epistemic interpretation would read, "It is impossible to know precise, simultaneous values of the position and the momentum of an electron."

The subtlety of Bohr's view becomes apparent when we consider his treatment of quantum systems when they are isolated. The epistemic view seems to imply that we cannot say anything about such systems since, being isolated, they are inaccessible to our knowledge. This is not, however, Bohr's view. Bohr concedes the human need to have a description, to have "the right words." In emphasizing that the description of an isolated system is "abstract," however, he disavows the ontic notion that such a description is a concrete representation of how nature really is. "There is no quantum world," he said. "There is only an abstract quantum description."[8]

7. Atoms Are Not "Little Things"

The history of atomism spans some twenty-five centuries. From Leucippus in the fifth century B.C. to John Dalton in the nineteenth century A.D., many scientists made discoveries and offered ideas about the nature of atoms. All these ideas, however, are based on the same premise, a premise so obvious that it is not even stated: *Atoms are little things.* They are similar to big things. The main difference between them and the things we see and handle is their size. The atomic theory went through many phases over the centuries, but this premise remained intact—until Bohr put it into question.

If atoms are "little things," they can be correctly described as substances, bits of matter, having their own attributes. They may not have all the attributes of "big things," but they do have the most basic attributes (Locke's "primary attributes"): size, shape, location, velocity, and, in addition, impenetrability. If we wish, we can visualize an atom and see it in our mind's eye, more or less correctly.

This idea is precisely what the framework of complementarity denies. According to Bohr, atoms are not "little things." They do not have attributes of their own. They show various attributes in various experimental set-ups; but *the attributes belong to the set-up as much as they belong to the atoms;* the attributes belong to what Bohr calls "the whole phenomenon."

The principle that the substance-attributes description has to be given up holds, of course, not only for atoms, but also for electrons, protons, molecules—all systems and entities in the atomic and subatomic domain. This conceptual break-through led to the resolution of the wave-particle duality. If a quantum system has no attributes of its own, then its apparent behavior as a wave phenomenon in one context and a particle in another is no longer untenable.

If we accept the idea of complementarity in relation to the atomic domain, we can't help wondering: What are the limits of its validity? How about systems that are larger than atoms? Where is the demarcation line between the quantum domain and everyday life? What about large molecules, are they "quantum systems" or "ordinary things?" What about viruses? What about bacteria?

Well, there is no demarcation line; but remember, momentum is mass times velocity; when we are dealing with a particle whose mass is much greater than the mass of an atom, a fairly large uncertainty in momentum translates to a very small uncertainty in velocity. When we measure the position of a billiard ball as accurately as we can, the ensuing uncertainty in its velocity is so small that it cannot be detected even with the most sensitive instruments. In principle, however, the idea of complementarity says something about all levels of nature, not just the atomic domain. The "thingness" of all objects, even tables and chairs, i.e., their substance-attributes description, is, in principle, merely an approximation; it disappears upon close scrutiny. If one wishes to describe any object accurately, one will eventually reach the point of realizing that its attributes do not belong to the object alone but to "the whole phenomenon" (i.e., object plus measuring apparatus).

Consider, for example, an ordinary table. It seems to have its own unique shape. But what happens when we wish to determine this shape very precisely? We have to find the precise boundary between the tabletop and the space above it. To do this we need to measure the location of atoms on the surface of the tabletop; the complementary features of these atoms will translate to complementary features regarding the shape, location, and momentum of the whole table.

So far we have found out what quantum systems are not: They are not "little things." Well, then, if they are not little bits of matter with attributes of their own, what *are* they?

Bohr loved to talk; and yet, throughout his long career, he was consistently silent on this issue. He stuck to his epistemic stance. Apparently the notion that quantum systems are systems which require complementary descriptions was all he needed. The question of the ontic vs. epistemic interpretation of quantum descriptions turns out to be very subtle indeed. We will come back to it repeatedly as we go along.

In the next chapter we introduce Erwin Schrödinger's "wave mechanics," a theory we mentioned in passing in Section 3, and Heisenberg's reading of it. The reading yields, provisionally, an ontic interpretation that is in line with the thrust of the framework of complementarity.

4. Waves of Nothingness

Erwin Schrödinger's discovery of "wave mechanics" led to insights
concerning the nature of electrons and other subatomic entities. When left
alone, electrons are not "things." They do not actually exist in space and
time; their existence is merely potential. They emerge into momentary
actual existence by acts of measurement. Hence, unlike classical
measurements, quantum measurements are creative; they literally create
the entities that are measured.

~

... and nothing is
But what is not.
　　　　　—W. Shakespeare

1. Schrödinger's Wave Mechanics

The University of Zurich was not a prominent center for theoretical physics during the 1920s, but Erwin Schrödinger, who had occupied the chair of theoretical
physics since 1922, kept abreast of the exciting new developments in the field. On
November 23, 1925, he gave a colloquium talk on de Broglie's thesis—the suggestion that the wave-particle duality applies not only to light but also to electrons.
Pieter Debye, a colleague of Schrödinger's who was in the audience, remarked
that he thought that "this way of talking was rather childish . . . to deal properly
with waves one had to have a wave equation." A month later Schrödinger left for
a Christmas vacation in the Alps. When he came back, he gave another lecture
and started by saying, "My colleague Debye suggested that one should have a wave
equation; well, I have found one."[1]

Schrödinger was not bereft of a sense of humor. He knew that the equation he
introduced so casually was one of the greatest discoveries of his time. Unlike Bohr
and Heisenberg, who spent months and years in torturous search, Schrödinger's
achievement was the result of a capacity for extreme concentration during a few
crucial weeks of a creative thrust.

But what is a wave equation, and why do we need one?

A wave equation is a concise way of expressing, mathematically, how the shape
of a wave changes from one moment to the next. Suppose we have "a wave func-

41

tion," a mathematical formula that describes the shape of a wave *at a given moment.* How will this shape change as time goes by? This is where the wave equation comes in: By solving it we can deduce, from the characteristics of a wave at one time, its future characteristics (and, if one wishes, its past characteristics as well) for all times.

Schrödinger arrived at his equation by putting together a known wave equation that was used in optics, where the wave behavior of light had been described mathematically for decades, and a formulation of mechanics that was introduced by Sir William Rowan Hamilton in the 1830s. Schrödinger's achievement was no less than the conversion of de Broglie's idea from the status of an enticing possibility to a full-fledged theory.

So, how does one check whether a wave equation is right? Schrödinger proceeded to calculate, on the basis of his equation, the properties of hydrogen atoms; his calculations agreed with the experimental results. *They also agreed with the results produced by "the new quantum mechanics" (also called "matrix mechanics") introduced by Heisenberg, Born, and Jordan more than a year earlier* (see Chapter 1, Section 1). This seemed strange, because the Heisenberg-Born-Jordan formalism was very different: It involved the manipulation of "matrices," i.e., large arrays of numbers, and the concept of "waves" had no place in it. Within a few months Schrödinger proved that, despite appearances, his wave equation and the Heisenberg-Born-Jordan formalism are mathematically equivalent. This result seemed hopeful indeed. As we saw in Chapter 1, "the new quantum mechanics" was merely a way of calculating the results of experiments. It came without a conceptual framework, without an explanation of what was really going on. It seemed as if Schrödinger's approach would provide this missing element—a clear description of the atomic phenomena in terms of waves.

But nature loves to hide; she refused to yield her secrets *that* easily. When Schrödinger examined the waves that were supposed to describe a free electron, an electron moving merrily, alone, in empty space, he found, to his dismay, that the waves dissipated. They became, in very short order, spread over a large region of space, much larger than the dimensions one could reasonably associate with an electron. He thus found himself in the same situation with which Heisenberg had to contend a year earlier, when he presented "the new quantum mechanics" at the colloquium in Berlin: The theory worked mathematically, but its meaning remained elusive.

Schrödinger's search for an interpretation of his wave function was based on the belief that the waves were as "real" as waves in water; he considered them to be "matter waves," actually existing and propagating continuously in space. Bohr was intrigued by Schrödinger's discovery but was, nevertheless, convinced that such an interpretation was impossible: Electrons, orbiting around the nucleus of an atom, seem to "jump" from one orbit to another, occupying one orbit, then, suddenly, occupying another. They are never found in the space between the orbits. This feature, which Bohr introduced in 1913 as an integral part of his model of the atom, was a cornerstone of quantum behavior. In Bohr's mind there was no way such a discontinuous behavior could be accounted for by continuously changing "matter waves."

The correct interpretation of Schrödinger's wave function was finally figured out by Max Born. In July 1926 Born published a paper in which he suggested that the square of the wave function indicates the *probabilities* for finding the particle at different locations. For example, if the value of the wave function of an electron at location A is *twice* its value at another location B, and one carries out a measurement to find out where the electron is, the probability of finding it at location A is *four times* greater than the probability of finding it at location B. (To be mathematically correct one has to consider "the absolute value squared" rather than the square, since the values of wave functions are, generally, complex numbers—but if you are not acquainted with complex numbers, don't worry about that.)

Bohr correctly saw Born's suggestion as a step in the right direction. But Schrödinger himself would have none of it. He firmly believed that his waves were material waves, not "probability waves." And in 1926 neither Bohr nor Schrödinger could suggest a consistent conceptual framework that would account for both "quantum jumps" and continuous wave behavior: Schrödinger's discovery of his equation as well as Born's probabilistic interpretation preceded Heisenberg's discovery of the uncertainty principle and Bohr's presentation of the framework of complementarity by more than a year.

Bohr invited Schrödinger to visit him in Copenhagen and discuss the issue. The visit took place in October 1926. The creation of the quantum theory, a process that started with Planck's hypothesis in 1900, was approaching a climactic conclusion, and that was reflected in the intensity of the discussions. Heisenberg, who was present, described them as follows:

> Bohr's discussions with Schrödinger began at the railway station and were continued daily from early morning until late at night. Schrödinger stayed at Bohr's house so that nothing would interrupt the conversations. And although Bohr was normally most considerate and friendly in his dealings with people, he now struck me as an almost remorseless fanatic, one who was not prepared to make the least concession or grant that he could ever be mistaken. It is hardly possible to convey just how passionate the discussions were, just how deeply rooted the convictions of each, a fact that marked their every utterance.
>
> . . . After a few days Schrödinger fell ill, perhaps as a result of his enormous effort; at any rate, he was forced to keep to his bed with a feverish cold. While Mrs. Bohr nursed him and brought in tea and cake, Niels Bohr kept sitting on the edge of the bed talking to Schrödinger: "But you must surely admit that . . ."[2]

In exasperation, Schrödinger said at one point, "If all this damned quantum jumping were really here to stay, I should be sorry I ever got involved with quantum theory."[3] Both Bohr and Schrödinger were groping in the dark, finding faults with each other's point of view regarding the quantum formalism, without, however, being able to suggest their own coherent interpretation.

2. The Wave and Particle Aspects in a Single Set-up:
The Double-Slit Experiment

Schrödinger's objections notwithstanding, Born's insight resolved one issue—the use of wave functions to predict the results of measurements. In doing so, however, it deepened the mystery of the nature of quantum systems. Let us look anew at the wave-particle duality, this time in the light of Born's probabilitics interpretation. In Chapter 3 we saw that the wave aspect and the particle aspect show up in different situations. According to the framework of complementarity they cannot show up in the same setting *whenever the measurement applies to an individual electron, a single quantum event*. If, however, one investigates the probabilistic or statistical properties of a quantum system by repeating the same measurement many times, applying it to many electrons, the wave and particle aspects can show up in a single setting. The setting we will look at will be the one with which we are familiar from our study of light; we will investigate the passage of electrons through narrow slits.

Consider the following experimental set-up (see Figure 4.1): A source, called an "electron gun," shoots a beam of electrons toward an opaque barrier with two very narrow slits, close together. Beyond the barrier there is a monitoring device, such as a TV screen or a photographic plate. We watch this TV screen to see how the arrival of electrons is registered.

As Figure 4.1 shows, the electrons arriving at the screen create a typical double-slit interference pattern. Such a pattern is characteristic of waves; it cannot be

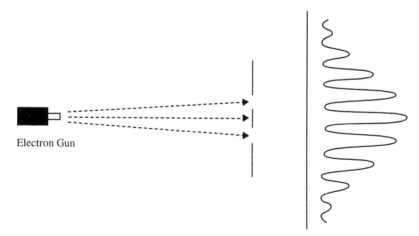

Figure 4.1   The double-slit interference experiment with electrons. The wavy line on the right represents the brightness distribution of dark and light spots on the screen.

explained in terms of particles. It seems, therefore, that our experiment with electrons demonstrates their wave nature.

But wait! The observed pattern was produced using a beam containing many, many electrons. What happens if we try to watch the arrival of electrons at the screen *one by one?* It seems as though we should finally be able to pin our electrons down, catch them by the throat, yell at them, "Are you particles or waves?!" and get an answer. If an electron is a wave, we can expect each electron to produce the whole interference pattern on the screen. To be sure, the pattern due to a single electron would be extremely weak—after all, the energy contents of a single electron are miniscule indeed—so that we may not be able to see anything; but as more electrons keep arriving, one by one, we should see the whole pattern emerging, hardly visible at first, then getting clearer and clearer.

But this is not what happens. Instead, each and every electron that gets through the slits makes a dot at some point on the screen, as if it were a particle. It is as if a single electron demonstrates the particle aspect, while the combined behavior of many electrons demonstrates the wave aspect.

Let's take a closer look (see Figure 4.2). We are watching the screen as the electrons begin to arrive, one at a time. What do we see? At first we just see some dots, apparently at random locations on the screen. By the time there are close to a hun-

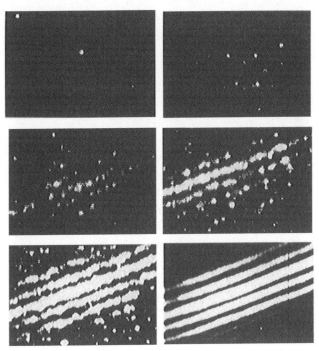

Figure 4.2   The way the screen of the double-slit interference experiment with electron looks in successive stages, as more and more electrons arrive at the screen. These pictures appeared in an article by P. G. Merli, G. F. Missiroli, and G. Pozzi, *American Journal of Physics* 66 (1976), p. 306. They are reproduced here courtesy of the authors.

dred electrons, however, the interference pattern begins to emerge: The incoming electrons seem to arrange themselves both randomly and in accordance with a pattern! Each single electron behaves unpredictably; we cannot tell where it will land, but the more electrons arrive, the more clear and distinct the interference pattern becomes!

This combination of order and randomness is characteristic of statistical events: When you flip a coin, you can't tell whether it will land heads or tails. But if you flip a thousand coins, about half of them will land heads, and the other half tails. Each coin behaves randomly *and* is subject to *a probability distribution* of equal probabilities for heads and tails.

The case of our interference pattern is similar, except that the probability distribution is more complex. The wavy line of Figure 4.1 indicates the relative probabilities of an electron arriving at the different locations. When a large number of electrons hit the screen, these relative probabilities become relative intensities: Locations of high probability are places where many electrons end up, hence the markings are intense, and in locations of low probability only a few electrons end up, hence the markings are weak.

Well, we tried to pin the electrons down and get an answer to the question "What are you?" and failed. We are left with a puzzle. When an electron hits the screen, it is localized like a particle. But the choice of its location is subject to a probability distribution that is wave-like! So—when the word "electron" comes up, how should we think about it? What concept of an electron would encompass both aspects of behavior?

3. What Is an Electron?

We are ready now to explore a line of thinking that will lead to a new concept of what an electron is. This new concept is, essentially, an expression of Heisenberg's ontic interpretation of quantum mechanics.

In facing the question "What is an electron?" we are essentially asking, "What is matter?" After all, an electron is just a small bit of matter, and all of the strange aspects of its behavior are shared by protons and neutrons, the other building blocks of the ordinary matter that makes up things like tables and chairs. We keep speaking about electrons merely to keep the presentation clear by focusing our attention on one example of a quantum system.

What, then, is an electron?

Well, an electron is an item (I am using the word "item" because it is the most neutral word I could find, a word that in itself does not imply anything about the nature of that to which it refers) that, in the set-up of the double-slit experiment, gives rise to a dot on the screen, the location of which is subject to a wave-like probability distribution. The probability distribution is decidedly "wave-like," because it is obtained through the principle of superposition, i.e., through the interference of the two waves that come from the two slits. Such a distribution cannot be obtained on the basis of a particle model, because particles are not subject to the principle of superposition.

Can we come up with a concept of an electron that will encompass both the particle-like dot on the screen and the wave-like probability distribution?

What happens when one tries to "imagine" an electron, that is, to have an image of it in one's mind's eye? The image starts out, perhaps, as a tiny hard ball; then, remembering the wave aspect, one may feel uncomfortable with this hard ball and try to make it fuzzy or add a cloud around it. After all, our ingrained concept of "matter" is "a collection of little hard objects"; if we didn't have to contend with the annoying wave aspect, the image of a tiny ball would have been right. According to the idea of complementarity, however, there is no appropriate image. If one wishes to work with images at all, one needs to evoke at least two complementary images. So maybe the challenge is to arrive at an appropriate *concept* rather than at appropriate *images*.

Can we have a *concept* of an electron without having any image? Yes, we can. Our powers of conceptualizing are greater than our powers of imagining. Most of our thinking is done, in fact, with concepts rather than with images. We think about beauty, justice, and equality; we think of happiness and love; we think about the number one million—all without images. But when we come to an electron or a proton, a bit of matter, we are conditioned to look for an image.

The source of the conditioning is not hard to find. It goes back to Newton, and to other atomists before him. We quoted Newton's concept of matter in Section 3 of Chapter 2:

> It seems possible to me that God in the beginning formed matter in solid, massy, hard, impenetrable, moving particles, of such sizes and figures, and with such properties, and in such proportion in space, as most conduced to the end for which he formed them.

Newton's paradigm has become the paradigm that informs us. But do we have an alternative?

Two thousand years before Newton, Aristotle wrote: "All things produced either by nature or by art have matter; for each of them is capable both of being and of not being, and this capacity is the matter in each." Dealing with a different issue, he commented, "One element is matter and another is form and one is potentially and the other actually."[4] Aristotle speaks of matter in terms of a capacity or a potentiality. This concept of matter is very different from Newton's "massy, hard, impenetrable, moving particles." Could it be relevant to our inquiry?

4. Electrons as Fields of Potentialities

Consider the idea of matter as a potentiality, or, rather, a collection or "field" of potentialities. To grasp what this means, let us begin with a close look at the concept of "potentiality."

When we say that something has a potentiality for being X, we are implying, first of all, that it is *not* X. For example, if Tommy's mother tells me that Tommy has the potential to be a great composer, she is clearly implying that, whether or not he will be a great composer someday, he is not a great composer now. Analogously, if we say that an electron is a collection, or a field, of potentialities, we are saying that it is not an actuality—that it does not exist at all as a "little thing." But this is clearly unacceptable! After all, the little dot on the TV screen, or on a photographic plate, clearly exists as a record of the electron's actual existence at that location. This dot can surely be considered a measurement of the location of the electron!

We are now getting closer to the heart of the matter. When Heisenberg was struggling toward the discovery of the uncertainty principle he realized, after weeks of groping in the dark, that the appearance of droplets in the cloud chamber pictures did not imply the existence of electron trajectories. The droplets showed that an electron existed at the location of the droplet; they did not imply that it existed between one droplet and the next. In the same spirit, let us ask the following question: Granted that the electron exists as an actual "thing" at the electron gun, where its departure is registered, and at the screen, where its existence gives rise to a visible dot, is there any reason to assume that it exists as an actual "thing" anywhere else?

According to Heisenberg's ontic interpretation, when the electron is not measured—in the space between the electron gun and the screen, for example— *it does not exist at all as an actual "thing." It exists merely as a field of potentialities.* Potentialities for what? For becoming an actual "thing," having certain properties, *if measured.* But as long as it is not measured, there is no thing there. We are suggesting that the electron actually exists at the electron gun and that it actually exists at the TV screen, but *it does not exist in between, except as a collection of potentialities.*

The term "measurement" was used in the previous paragraph in a generalized sense: Any interaction that involves the electron and could be used, in principle, to learn anything about it counts as a "measurement." Thus any interaction of the electron with bits of matter that are much larger and more complex than atoms or molecules can be considered a measurement.

Heisenberg's view naturally resolves the "wave-particle duality" that troubled physicists for decades: electrons are waves when they are fields of potentiality, and particles—when their existence is actual!

Heisenberg's ontic interpretation introduces a veritable paradigm shift. It sounds strange; it makes us feel uneasy—until we get it. The thrust of the idea can be explained in terms of a mundane example, "the refrigerator door effect": Whenever we open the refrigerator door, we see the inside light turned on. A small child tends to assume that the light inside the refrigerator is always on. You may even remember the revelatory experience you had when, as a child, you understood for the first time that the inside light is *not* always on! We may be, in relation to the existence and non-existence of electrons, in a similar position: Whenever we measure them, they are there. Our assumption that they continue to be there even when left alone is analogous to the small child's belief regarding the refrigerator light.

When an electron is not being measured, it is nowhere is space, and yet, if its position is measured, it will be found somewhere. If its momentum is measured, some definition value of the momentum will be found. It *follows that it is the measurement that brings it into actual existence.* There is a profound difference between ordinary measurements and quantum measurements. In ordinary life, and in classical physics, a measurement gives information about the state of the measured system, a state that is not significantly affected by the measurement process. In quantum physics, however, *measurements are creative*: They literally create the electron as an actual thing, where, before the measurement, no thing existed.

This new concept of the electron brings new meaning to the relationship between the particle-like and wave-like aspects of the two-slit interference experiment: The wave-like behavior describes the field of potentialities, and the particle-like behavior describes the result of a position measurement. When its position is measured, the electron ceases to be merely potential—it becomes actual. The complementarity of particles and waves translates into interplay between two modes of being, the potential and the actual. But how do measurements bring about this mysterious transition from the potential to the actual? A detailed study of this question will be taken up in Chapters 10, 11, and 16.

5. Description of Reality or Description of Knowledge?

In the closing remarks of the last chapter I wrote that Heisenberg's ontic interpretation is in line with the thrust of the idea of complementarity. This is a strange statement. The framework of complementarity demands two different descriptions; how can one ontic description fit the bill?

Well, Heisenberg's ontic interpretation is not your run-of-the-mill ontic interpretation. In response to the question "What is an electron?" it says, in essence, *"An electron is not."* More precisely: "An electron does not exist in actuality, except when measured; when not measured, it is a collection of potentialities." The secret of the strange conformity between this ontic description and the demands of complementarity lies in the distinction between actual existence and potential existence: A collection of potentialities can have complementary and mutually exclusive characteristics; an actual entity cannot. For example, an electron can have the *potential* for being simultaneously at two different locations. This means that before being measured it has the possibility of being at location A as well as the possibility of being at location B. When it is *actually* measured, however, it will be found either in location A or in location B, not in both. We will come back to the distinction between potentialities and actualities in Chapter 16.

Incidentally, although Heisenberg introduced an ontic interpretation of Schrödinger's waves which has become mainstream,[5] his own position regarding the question of the ontic interpretation versus the epistemic of the quantum theory is rather complex. The ontic interpretation is introduced in the following paragraph:

> The probability function combines objective and subjective elements. It contains statements about possibilities or better tendencies ("potentia" in Aristotelian philosophy), and these statements are completely objective, they do not depend on any observer; and it contains statements about our knowledge of the system, which of course are subjective in so far as they may be different for different observers. In ideal cases the subjective element in the probability function may be practically negligible as compared with the objective one.[6]

However, in at least one essay, written in 1958, he advocates an epistemic interpretation of "the laws of nature that we formulate mathematically in quantum theory."[7]

While Heisenberg's concept of quantum systems as field of potentialities is in line with the framework of complementarity, a profound disparity remains. Heisenberg's interpretation of quantum systems provides a unified and consistent description. It is ontic rather than epistemic: It tells us what quantum systems are. As we saw in the last chapter, however, Bohr refused to accept an ontic inter-pretation. From his point of view, since the two complementary descriptions are all "we can say about nature," nothing more is needed. Which interpretation is correct, the ontic or the epistemic?

Heisenberg's equivocation indicates that something is missing from both the ontic and epistemic interpretations. In Chapter 16, Section 5 we will finally resolve the issue. Purely as a matter of convenience, we will adopt, provisionally, the ontic mode of description, because its language is simpler and more direct. Take the complementarity of position and momentum as an example. Its ontic expression is, "At any given moment, the more well-defined the position of a quantum system, the more ill-defined its momentum." By contrast, the epistemic equivalent is rather cumbersome: "If simultaneous measurements are carried out on a quantum system, the more well-defined the result of the position measurement, the more ill-defined the result of the momentum measurement."

Both Bohr's and Heisenberg's interpretations of the quantum theory point toward a new emergent paradigm concerning the nature and knowledge of the physical world. By the fall of 1927 the quantum theory, the basis of the paradigm, reached maturity. But Albert Einstein, the greatest physicist of the century, was not buying either interpretation. He understood and appreciated the successes of quantum mechanics but steadfastly refused to accept it as a fundamental theory of nature. As we will see in Chapter 6, the brilliant arguments he discovered in support of his position could not be easily dismissed.

5. Paul Dirac and the Spin of the Electron

The purpose of this chapter is to introduce the reader to the last of the founding fathers of quantum mechanics, Paul Adrien Maurice Dirac, and to the concept of "the spin of the electron," a concept which will be crucial in the discussion of Bell's experiment in Chapter 7.

~

It seems to me that one of the fundamental features of nature is that fundamental physical laws are described in terms of a mathematical theory of great beauty and power.

—Paul Dirac

1. Dirac's Transformation Theory

Some of the great physicists we encountered in the previous chapters—Bohr and Heisenberg are prime examples—not only liked to talk but needed talking as an ingredient in the process that led to their discoveries. Heisenberg went as far as writing, "science is rooted in conversations."[1] And then there is the other type of physicist, of which Einstein is a prime example: the loner. The loner seems to spend most of his or her creative time away from words, exploring depths of thought that are essentially non-verbal; the use of words is not a part of his or her creative process. Words are used only to express and communicate the final results of the non-verbal thinking.

Paul Dirac was such a loner. His concise expressions sent many physicists, myself included, into explorations of the depths of the ideas of quantum mechanics. His adverse relationship to talking was rooted, to some extent, in his early childhood experiences. When he was sixty years old he told Thomas Kuhn, in an interview:

My father made the rule that I should only talk to him in French. He thought it would be good for me to learn French in that way. Since I found that I couldn't express myself in French, it was better for me to stay silent than to talk in English. So I became very silent at that time—that started very early.[2]

When Dirac visited Madison, Wisconsin, in 1929, the residents learned about him from a local newspaper interview that began as follows:

I been hearing about a fellow they have at the U. this spring—a mathematical physicist, or something, they call him—who is pushing Sir Isaac Newton, Einstein and all the others off the front page. . . . His name is Dirac and he is an Englishman. . . . So the other afternoon I knocks at the door of Dr. Dirac's office in Sterling Hall and a pleasant voice says "Come in." And I want to say here and now that this sentence "come in" was about the longest one emitted by the doctor during our interview.[3]

Dirac was a little younger than Heisenberg, and much younger than Bohr and Schrödinger. Studying on his own, in Cambridge, England, he quickly assimilated "the new quantum mechanics" introduced by Born, Heisenberg, and Jordan, as well as Schrödinger's "wave mechanics." Then, while working at Bohr's institute in Copenhagen, he came up with a new, abstract, and deep formulation of the quantum theory, called "transformation theory." Dirac's formulation clarified the relationship between classical physics and quantum mechanics and, in a sense, put the new quantum mechanics and wave mechanics in proper perspectives.

I must confess that in my own heart I have a soft spot for Dirac. When I was a junior in college and had developed the wish to *really* understand quantum mechanics, I found out that Dirac's book *The Principles of Quantum Mechanics* was considered the most profound and most difficult exposition of the subject. I plunged into it with enthusiasm and perseverance, doing little else for a few weeks, sometimes spending hours pondering a single paragraph. The experience was not a disappointing one. Dirac's precise, pithy formulations led me to satisfying experiences of deep understanding, and to feelings of gratitude to the man who made them possible.

Transformation theory was Dirac's first major achievement. It was followed by a series of important contributions. He arrived at his most famous discovery in early 1928: a reconciliation of the budding quantum theory with Einstein's theory of Special Relativity—and this is where the spin of the electron comes into the picture.

2. The Physics of Spinning Objects

As we have seen, an electron is very different from the objects we encounter in everyday life. Not surprisingly, then, we will have to think of a spinning electron in a way that is very different from the way we think about, say, a spinning bicycle wheel. Nevertheless, our discussion of the spin of the electron must be grounded in an understanding of the properties of spinning objects in familiar, everyday conditions.

In Chapter 3, Section 4, we discussed the concept of "momentum." We saw that the momentum of an object is a measure of "how much motion" the object has; that momentum is a vector quantity (a quantity that has both magnitude and direction and can be represented by an arrow), and that it has components in the

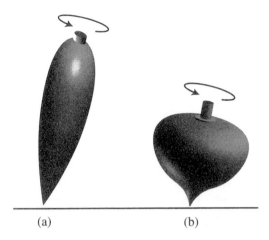

(a)   (b)

Figure 5.1   Comparing the angular momentum of top (a) with the angular momentum of top (b). If they have the same mass and spin at the same rate, top (b) will have more angular momentum than top (a).

directions of the x-, y-, and z-axes. We also saw that momentum is a conserved quantity: As long as the system is undisturbed, the sum of the momenta of its constituents stays the same, regardless of how they interact with each other.

A corresponding quantity, called "angular momentum," or "spin," can be defined for rotations. This quantity measures "how much rotation" an object, or a system of objects, has (see Figure 5.1). Just like momentum, the spin is a vector quantity, it has components along the x-, y-, and z-directions, and it can be represented by an arrow. The length of the arrow corresponds to "the amount of rotation" the object has, and its direction is determined by the axis of rotation (see Figures 5.2 and 5.3). Furthermore, spin, just like momentum, is conserved: As long as a system is left undisturbed, the sum of the angular momenta of its constituents stays the same.

To become more familiar with this new concept of spin, let us see now how the spins of different objects add up. Consider the following situation: A bicycle wheel rotates clockwise around a vertical axis, and another bicycle wheel counterclockwise around another vertical axis. The spin of each wheel is represented by a vector. The two vectors are of the same length, but one of them points down, corresponding to a clockwise rotation, and the other, which represents a counterclockwise rotation, points up. Now suppose we wish to consider these two particles to be one system (see Figure 5.4). What is the total spin of this system? Since the amount of rotation up is equal and opposite to the amount of rotation down, the two rotations cancel each other out. Each part of the system, that is, each wheel, does have a spin, but the two rotations add up to no spin. This is why an ice skater, standing perfectly still, having no angular momentum, can get her torso rotating to the right simply by swinging her arms to the left.

Figure 5.2   Representing angular momentum by a vector. The length of the arrow represents the magnitude of the vector, and the direction of the arrow represents the direction of the vector. The length of the arrow tells us "how much rotation" there is, while the direction of the arrow is the direction of the axis of rotation.

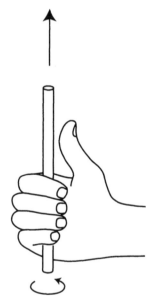

Figure 5.3   The direction of the arrow tells us whether the rotation is clockwise or counterclockwise according to the following convention: When the fingers of the right hand curl around the rotation axis in the direction of the rotation, the thumb points in the direction of the arrow. According to this convention, clockwise and counterclockwise rotations correspond to arrows pointing in opposite directions.

Figure 5.4   To find the total angular momentum of a system composed of two particles, add up the angular momenta of the two particles. In the case depicted in the picture, since the amount of rotation up is equal and opposite to the amount of rotation down, the two rotations cancel each other. Each part of the system—that is, each particle—does have a rotation, but the two rotations add up to no rotation.

3. The Spin of the Electron

Returning to the quantum domain, let us consider an electron. Does an electron have a spin?

There is no *a priori* reason to assume that an electron has or does not have spin, but investigations of the properties of light emitted from atoms showed that electrons do have spins. Since electrons are not "little things," it would be wrong to picture them in our minds' eyes as "little spinning things." An electron does have spin, but this spin has two quantum characteristics that have no counterparts in the familiar world of spinning tops and bicycle wheels.

First, different components of the spin do not have well-defined values simultaneously. We mentioned in Section 2 that the spin vector can be decomposed into components, so that we can discuss the x-component, the y-component, and the z-component of the spin. It turns out, however, that when we are dealing with an electron, or with any other quantum system, it is impossible to carry out measurements that will yield the magnitude of more than one component: For example, if one measures accurately the z-component, one cannot measure simultaneously and accurately either the x-component or the y-component.

There is one exception to this rule: If all three components are zero, that is, if a quantum object has no spin, it is possible to ascertain that all three components are simultaneously zero. This does not apply to the electron, the spin of which has a fixed value which is not zero, but it does apply to some of the other elementary particles, as well as certain nuclei and certain atoms.

This exception notwithstanding, the relationship between the spin components of a system is analogous to the relationship between the position and the momentum. According to the uncertainty principle, if the position of an electron is measured precisely, its momentum cannot have, simultaneously, a well defined value. Similarly, if one of the spin components of the electron is well defined, the other two components do not have well-defined values.

The second quantum characteristic possessed by electrons' spin is that the spin is "quantized": For a rotating top, a measurement of the x-, or y-, or z-component of the spin can yield any value. For an electron, however, a measurement of a spin component—any spin component—always yields one of two "allowed" values. Some elementary particles, and many atoms, have more than two allowed values. They too are "quantized"; however, only a small number of values are allowed—three, four, five, sometimes as much as eleven, and in rare cases more. For electrons, protons, and neutrons only two are allowed. Any measurement of a spin component yields one of them.

Take, for example, a measurement of the x-component of the spin of an electron. The measurement always yields one of two allowed values that are equal in magnitude but opposite in direction. The positive one indicates spin in the direction of a right-handed screw along the x-axis and is called "spin up." The negative one indicates spin in the direction of a left-handed screw along the x-axis, and is called "spin down." In case you are curious about the numbers: Naturally, the numerical value of a measured spin component is very small when denoted in units designed for the measurement of everyday objects. In the system of units in which mass is measured in kilograms, length in meters, and time in seconds, the unit of spin is called "Joule-second." Rotating bicycle wheels have spins that are of the order of a few Joule-seconds. In contrast, the two allowed values of a component of the electron spin are either plus or minus five billion billion billion billionths of a Joule-second.

4. The Relativistic Wave Equation

Let us recall at this point the beginning of the previous chapter, in which Schrödinger's wave equation was introduced.

Schrödinger's equation was a great success. There was, however, something wrong with it: It did not fit in with Einstein's theory of Special Relativity. Had the equation been discovered before Einstein introduced his theory—that is, had quantum physics been developed in a Newtonian context—Schrödinger's equation would have been perfect. By the 1920s, however, Special Relativity was well established, and Newton's framework was understood as fundamentally flawed: It is applicable to particles moving slowly relative to the speed of light merely because, for such particles, its results are a good approximation to the exact results obtained from calculations using Special Relativity.

Schrödinger was well aware of this drawback of his equation. The first equation he wrote down was, in fact, a relativistic one. He discarded it, however, because of what he considered undesirable characteristics and limited himself to the non-relativistic approximation. The question "What is the correct wave equation that agrees with Special Relativity?" remained unresolved.

This is where Dirac came into the picture. Dirac realized that a mathematical description of an electron that agrees with Special Relativity calls for the use of more than one wave function. As a consequence, Schrödinger's equation would have to be replaced by more than one equation. At first he tried to introduce two wave functions and two equations, but that did not work. After considerable struggle he introduced four wave functions and a corresponding set of four equations. They seemed to fit the bill.

Dirac's four equations are referred to as "Dirac's equation" because the four wave functions are usually written as one column. The column is thought of as one mathematical quantity, called "spinor," having four "components," just as when a vector is expressed in terms of its three components, it is still thought of as one mathematical entity, one vector.

Any doubts Dirac might have had about the correctness of his equation were dispelled when he discovered that the equation describes a spinning electron, the properties of which agree with the characteristics that were observed experimentally.

As mentioned above, the present chapter is an interlude. Let us rejoin the main story.

6. An Irresistible Force Meets an Immovable Rock

Bohr and Einstein respected and loved each other while being in complete disagreement about the way quantum mechanics should be interpreted. Einstein's rejection of quantum mechanics as a complete, fundamental theory went through two phases. In the first phase he proposed "thought experiments" that were supposed to invalidate the uncertainty principle. When this failed, he and his assistants Podolsky and Rosen proposed a thought experiment which claimed to prove that quantum mechanics is incomplete. The argument failed to convince Bohr.

~

Quantum mechanics demands serious attention. But an inner voice tells me that this is not the true Jacob. The theory accomplishes a lot, but it does not bring us closer to the secrets of the Old One.

—Albert Einstein

1. A Clash of Minds

The physicist and author Abraham Pais, who knew both Einstein and Bohr, wrote about the first time that he witnessed a conversation between them. The setting was the Princeton Bicentennial Meetings in 1946; this was also the occasion at which Pais met Einstein for the first time:

> I missed the first opportunity to catch a glimpse of Einstein as he walked next to President Truman in the academic parade. However, shortly afterwards Bohr introduced me to Einstein, who greeted a rather awed young man in a very friendly way. The conversation on this occasion soon turned to the quantum theory. I listened as the two of them argued. I recall no detail but remember distinctly my first impressions: they liked and respected each other. With a fair amount of passion, they were talking past each other.[1]

The Bohr-Einstein controversy is, perhaps, the greatest intellectual debate of the twentieth century. It spanned nearly thirty years, starting at the Fifth Solvay Conference in Brussels in October 1927 and ending, unresolved, with Einstein's death in 1955. In some sense it did not even end then: After Einstein's death Bohr continued to argue with him, in his mind, "as if Einstein were still alive."[2]

Studying the Bohr-Einstein controversy, I am awed by the depth of the thoughts produced by these great minds, as they rose to the highest levels of subtlety and clarity in their efforts to meet each other's challenges. This study has another attractive feature: One does not have to contend with the nasty manifestations of an ego clash that so often accompany intellectual debates. Throughout its duration, Bohr and Einstein maintained not only a high level of courtesy but also the greatest genuine respect, high regard, and love for each other.

2. The Beginning

The first meeting of Einstein and Bohr took place in 1920. By then the so-called early quantum theory, which included Bohr's model of the atom and Einstein's proposal that under certain conditions light travels like a particle, was well known to the major players in the unfolding drama—the birth of the quantum theory. And every one of these major players, Bohr and Einstein included, was trying to meet the challenge of creating a comprehensive theory that would accommodate the successful features of both classical physics and the budding quantum theory. Even in this first meeting Bohr and Einstein had their differences: Bohr refused to accept Einstein's idea that light can travel like a particle. It was not until the Fifth Solvay Conference in October 1927, however, that their debate regarding the meaning and significance of quantum mechanics was launched.

By the fall of 1927 all of the ingredients of the full-fledged quantum theory were in place: the Born-Heisenberg-Jordan "new quantum mechanics," Schrödinger's wave mechanics, Born's statistical interpretation of the wave function, Heisenberg's uncertainty principle, and Bohr's framework of complementarity.

During the public sessions of the Fifth Solvay Conference Einstein spoke only once. He started by saying, "I must apologize for not having penetrated quantum mechanics deeply enough,"[3] and proceeded to show an apparent contradiction between quantum mechanics and the theory of Special Relativity.

Einstein was well aware of the successes of the quantum theory in predicting the results of experiments, yet he felt uneasy about it. His struggle to give this feeling of unease a precise expression went through two distinct phases. In the first phase he tried to disprove the uncertainty principle by finding thought experiments that violate it. When this approach failed, he tried to prove that the quantum theory is incomplete, i.e., that there are properties of subatomic particles that it fails to represent. At the time of the Fifth Solvay Conference Einstein's struggle was still in its first phase. Adopting a straightforward line of attack, he was determined to prove that the uncertainty principle can be violated in physical experiments.

3. The Uncertainty Principle Under Fire

While Heisenberg was working on his uncertainty principle he discovered, in addition to the uncertainty relation involving position and momentum, another uncertainty relation, involving energy and time. This other principle puts a limit on the level of precision to which it is possible to measure simultaneously the energy of a system and the time to which this measurement refers. Einstein's attack on the uncertainty principle consisted of suggestions for experiments, some of which were supposed to violate the position-momentum uncertainty principle, while others were directed against the time-energy uncertainty principle. The beauty of his approach was that it was not necessary to *perform* the suggested experiments! To show that the uncertainty principle is invalid, it is enough to show that it is possible *in principle* to conduct measurements that violate one or the other of the uncertainty principles. The experiments discussed were *thought experiments*. They were limited only by the capacity of the creative imagination, not by the level of the available technology.

Although Einstein did not say much during the public sessions of the Fifth Solvay Conference, he was, privately, very active indeed. Here is Heisenberg's recollection of the conversations between Einstein and Bohr:

> The discussions usually started at breakfast, with Einstein serving us up with yet another imaginary experiment by which he thought he had definitely refuted the uncertainty principle. We would at once examine his fresh offering, and on the way to the conference hall, to which I generally accompanied Bohr and Einstein, we would clarify some of the points and discuss their relevance. Then, in the course of the day, we would have further discussions on the matter, and, as a rule, by suppertime we would have reached the point where Niels Bohr could prove to Einstein that even his latest experiment failed to shake the uncertainty principle. Einstein would look a bit worried, but by the next morning he was ready with a new imaginary experiment, more complicated than the last, and this time, so he avowed, bound to invalidate the uncertainty principle. This attempt would fare no better by evening.[4]

This game of challenge and rebuttal continued when Bohr and Einstein met again at the Sixth Solvay Conference in 1930. This time Einstein came up with an ingenious thought experiment, which, he believed, proved conclusively that the energy-time uncertainty principle can be violated. Here is how Leon Rosenfeld, who participated in the conference, described Bohr's reaction:

> It was quite a shock to Bohr . . . he did not see the solution at once. During the whole evening he was extremely unhappy, going from one to the other and trying to persuade them that it couldn't be true, that it would be the end of physics if Einstein were right; but he couldn't produce any refutation. I shall never forget the vision of the two antagonists leaving the club [of the *Fondation Universitaire*]: Einstein a tall majestic figure, walking quietly, with a somewhat ironical smile, and Bohr trotting near him, very excited. . . . The next morning came Bohr's triumph.[5]

Bohr spent a sleepless night, at the end of which he finally understood where Einstein went wrong: Einstein had forgotten to take into consideration a certain effect which is an integral part of Einstein's own theory of General Relativity!

There is no question that Bohr derived an immense amount of pleasure from his rebuttal, and especially from the irony of the source of Einstein's mistake. In Chapter 3 I mentioned having attended a lecture by Bohr in Israel in May 1958. The one item I remember from the lecture is a slide showing the apparatus of Einstein's thought experiment, Einstein's last challenge to the uncertainty principle. Twenty-eight years after the event, Bohr was still savoring his victory!

This episode marked Einstein's last challenge, not to the quantum theory, mind you, just to the uncertainty principle. Following his spectacular defeat, he accepted the idea that the limitations imposed by the uncertainty principle are valid, but he did not like it. The following incident was recounted by Abraham Pais, who heard about it from his friend Hendrik Casimir: When Einstein gave a colloquium in Leiden, Holland, on this last thought experiment, in November 1930, and someone made the comment that there is no conflict within quantum mechanics, Einstein responded by saying: "I know, this business is free of contradictions, yet in my view it contains a certain unreasonableness."[6]

Following this last challenge and rebuttal the ball was clearly in Einstein's court. It took him five more years to give his feelings of the "certain unreasonableness" of quantum mechanics the form of a precise scientific argument. But when he let the ball fly, his challenge to the idea that quantum mechanics is complete created quite a stir; it continues to reverberate in the physics community to this day.

4. Is Quantum Mechanics Incomplete?

The thrust of Einstein's efforts during the second phase of his debate with Bohr was to demonstrate that quantum mechanics is incomplete; and, being incomplete, it does not qualify as the new, complete, fundamental theory of nature. He compared quantum mechanics with statistical mechanics, the classical theory that deals with the statistical properties of a large number of particles. When statistical mechanics is applied, for example, to a volume of gas, it predicts its behavior very well: If the gas is about to gradually expand or contract, statistical mechanics can predict what changes will take place in its pressure and temperature. Statistical mechanics cannot give, however, the exact configuration of the gas molecules; it only gives probabilities for different configurations. And it cannot give more than probabilities not only because the number of molecules involved is so large that it is impossible to follow each one but also because the initial conditions of the system—the location and velocity of each and every molecule—are unknown. If it were possible to know the position and velocity of each and every molecule at the present moment, the laws of classical mechanics would determine, in principle, their future movement, and there would be no need to invoke probabilities. The use of probabilities in statistical mechanics is a reflection of our ignorance regarding the precise initial configuration of the molecules.

Einstein believed that quantum mechanics invokes probabilities for a similar reason and that one should try to discover the more fundamental theory that, like classical mechanics, does not invoke the concept of probability at all. Bohr believed, however, that there is no fundamental theory underlying quantum mechanics. He firmly believed that quantum mechanics was it.

Einstein's reservations about quantum mechanics were buttressed by his conviction that there is no place for probabilities in a theory that can give us, in principle, a complete description of any physical system; that, as he said repeatedly, "God does not play dice." But in spite of Einstein's enormous prestige, most physicists sided with Bohr. Could Einstein *prove* that the quantum theory was incomplete?

In order to prove that a theory is incomplete, it is necessary to show that there are aspects or properties of things or events, some "elements of reality," as Einstein called them, which are indubitably there and can be measured, yet cannot be accounted for within the framework of the theory. In discussing a concept like "an element of reality" we are on slippery ground: A full, precise definition of the term is tantamount, more or less, to a resolution of the deepest issues of philosophy! As we shall see, however, Einstein managed, in 1935, with the help of two young assistants, Boris Podolsky and Nathan Rosen, to circumvent the philosophical issues, use the concept of "elements of reality," and give it a precise, unambiguous meaning.

5. "Elements of Reality"

The uncertainty principle is embedded in the mathematical formalism of the quantum theory: Schrödinger's wave functions cannot describe a particle having simultaneously a well-defined position and a well-defined momentum. Therefore a proof that a particle *can* simultaneously have a well-defined position and a well-defined momentum is tantamount to a proof that the quantum theory is incomplete. If the mathematical formalism of the theory does not contain a wave function that corresponds to such a state, it follows that the theory is incapable of describing a quantum state that actually exists; ergo, the quantum theory is incomplete.

The approach adopted by Einstein, Podolsky, and Rosen (to be referred to as EPR) in their 1935 paper, which appeared in *The Physical Review*, is different from Einstein's line of attack during the first phase of the debate. EPR did not suggest a way to measure, simultaneously, the position and momentum of a particle. Rather, they presented a proof that a well-defined momentum and a well-defined position of a particle can exist simultaneously, that both the position and the momentum of a particle are "elements of reality." But how can this assertion be proved? How does one prove that anything is "an element of reality"?

This question was tackled in the EPR paper in a remarkable fashion. The authors realized that while the question "What is an element of reality?" would get them, philosophically, into deep waters, their aim of proving that the quantum theory is incomplete does not necessitate a full answer to the question. What is needed is only a *sufficient condition* for an item to be an element of reality, a sufficient condition that need not be a *necessary condition*.

Let us explore these new concepts: What is a sufficient condition? What is a necessary condition? What is a necessary *and* sufficient condition? And what is a condition that is sufficient but not necessary, like the one EPR formulated for their "elements of reality"?

Suppose we are looking for conditions for a number to be a multiple of 9. A *sufficient* condition would be, for example, the condition that it is a multiple of 27: This condition is sufficient because if a number is a multiple of 27, then it is also a multiple of 9. This condition, however, is not necessary: There are plenty of numbers, e.g., 18 or 36, that are multiples of 9 without being multiples of 27. So, being a multiple of 27 is a sufficient but not necessary condition for a number to be a multiple of 9. Now let us look for a *necessary* condition. The condition that our number is a multiple of 3 fits the bill: All multiples of 9 are also multiples of 3, so a number that is a multiple of 9 is *necessarily* a multiple of 3. But this condition is not *sufficient*: There are many numbers, e.g., 15 or 21, that are multiples of 3 without being multiples of 9.

In the previous paragraph we have constructed a condition that is sufficient but not necessary and another one that is necessary but not sufficient. Can we find a condition that is *both necessary and sufficient* for a number to be a multiple of 9? Here is a suggestion: A necessary and sufficient condition for a number to be a multiple of 9 is that the sum of its digits is a multiple of 9. We claim that if a number is a multiple of 9, the sum of its digits is a multiple of 9. Conversely, if the sum of its digits is a multiple of 9, then the number itself is a multiple of 9. For example, 846 = 94 x 9 is a multiple of 9, and so is the sum of its digits, 8 + 4 + 6 = 18. 1914 = 212 x 9 + 6 is not a multiple of 9, and neither is the sum of its digits, 1 + 9 + 1 + 4 = 15. This condition is infallible and is a quick and practical way to determine whether a large number is or is not a multiple of 9, without having to go to the trouble finding out whether it divides by 9 without a remainder.

EPR have figured out, as a first step in their argument, that they did not need to tackle the difficult task of finding a necessary and sufficient condition for something to be an "element of reality." It was enough for them to find a condition that is sufficient but not at all necessary—a condition that will lead everyone to say, "Well, if this condition is satisfied, then of course this property of a particle is a property that exists in reality."

The sufficient condition they chose is the following: "If, without in any way disturbing a system, we can predict with certainty (i.e., with probability equal to unity) the value of a physical quantity, then there exists an element of physical reality corresponding to this physical quantity."[7] Of course, this is not a *necessary* condition; there are many elements of reality that exist in systems that *are* disturbed, but these were of no interest to EPR. They chose a sufficient condition of a special kind, referring to systems that are not disturbed, and they had their reason for doing so, as we shall see.

What is the meaning of the proposed sufficient condition? Consider a mundane example: The other day I handed my friend Amy ten dollars in one-dollar bills and two boxes, a red one and a blue one. She put some of the dollar bills in the red box and the remaining bills in the blue box. She then put the red box on a high shelf and left the blue box on the table. I haven't noticed how many bills she put in the red box, and I wished to know. Being lazy, I didn't want to climb on a chair to reach the red box. And there was no need for me to do so. All I needed to do was open the blue box and count the number of bills it contains. If I find six dollar bills, I can be sure that the presence of four dollar bills in the red box is an element of reality.

As this example demonstrates, EPR's sufficient condition is very simple. Yet, as we will see presently, they got a lot of mileage out of it.

6. The EPR Thought Experiment

The thought experiment proposed by EPR is this: Consider two particles, particle A and particle B, moving to the right and left, respectively. According to Heisenberg's uncertainty principle it is impossible to measure, simultaneously, the precise position and precise momentum of either particle. However, for the system composed of these two particles, quantum mechanics allows simultaneous measurement at any given moment t_0 of the precise values of (1) the total momentum of the system, i.e., the sum of the momenta of the two particles, and (2) the distance between them. Furthermore, according to quantum mechanics it is possible to arrange things so that at any given moment t_0 the total momentum of the system will be zero and the distance between the particles will be any chosen quantity; call it x_0. The fact that the total momentum is zero means that the momenta of the two particles are equal in magnitude and opposite in direction.

Now let us ask a question: Is the position of particle B at the given time t_0 an element of reality? (That is, does B occupy some precise position at that time?) To answer the question, we need to refer back to the sufficient condition for a property to be considered an "element of reality" and ask: Can we find the position of B at time t_0 without disturbing it in any way? The answer is: Yes, we can. We know the distance x_0 between the two particles. If we measure the position of A at that time, we can calculate the position of B without having disturbed it.

This method is similar to the one I used when I opened the blue box to find out how many dollar bills were in the red box. Since the distance between particles A and B is a known quantity, I don't need to disturb B to find where it is. I can carry out my measurement on A and deduce where B must be. We have found a way to predict with certainty B's position without disturbing it in any way. Conclusion: The position of particle B is an element of reality.

Now let us ask a different question: Is the *momentum* of B an element of reality? Can we find this momentum without disturbing B? Once again, the answer is yes. The momentum of A as it travels to the right is equal in its magnitude to the momentum of B as it travels to the left. Hence we can measure the momentum of A and deduce what the momentum of B must be. Conclusion: The momentum of particle B is an element of reality.

EPR thus claimed to have proved that both the position and the momentum of a particle, particle B, are elements of reality; that is, a particle can have simultaneously a well-defined position and a well-defined momentum. Quantum mechanics cannot describe such a particle; ergo, quantum mechanics is incomplete.

This is the EPR argument. What was Bohr's response? Let us quote Leon Rosenfeld again (he was in Copenhagen at the time):

> This onslaught came down upon us as a bolt from the blue. . . . As soon as Bohr heard my report of Einstein's argument, everything else was abandoned: we had to clear up such a misunderstanding at once. . . . In great excitement, Bohr immediately started dictating to me the outline of . . . a reply. Very soon, however, he became hesitant: 'No, that won't do, we must try all over again . . . we must make it quite clear . . .' So it went on for a while, with growing wonder at the unexpected subtlety of the argument. . . . The next morning he at once took up the dictation again, and I was struck by the change in the tone of the sentences; there was no trace in them of the previous day's sharp expressions of dissent.[8]

Bohr's rebuttal appeared in *The Physical Review* three months after the EPR paper. His position was that the two particles, A and B, cannot be considered two systems. As long as no measurements are performed, they are "entangled"—they constitute one system. Moreover, this system is different according to which measuring apparatus it interacts with. The situation of being subjected to an apparatus that measures the position of A is complementary to the situation of being subjected to an apparatus that measures the momentum of A. Hence when this system is subjected to a measurement of position, its reality is different from when it is subjected to a measurement of momentum.

This time, unlike the challenge and rebuttal of 1930, it was not so obvious who was right. Einstein did not accept Bohr's rebuttal. Bohr was convinced that he was right, while Einstein stood by his sufficient condition for an "element of reality."

There is a particular aspect of the EPR thought experiment that tends to give credence to Einstein's position. The measurement of momentum which is carried out on particle A in order to deduce the momentum of particle B can be carried out, in principle, while A and B are far away from each other, even light-years away! Einstein's position agrees with local realism: If the momentum of B is an element of reality, then such a measurement, carried out on particle A, does not affect B in any way. It merely lets us deduce the momentum that particle B had all along. Bohr's view, however, violates local realism. Bohr believed that, because particles A and B are entangled, subjecting A to a momentum-measuring apparatus puts B too, instantly, into a state of having a well-defined momentum. Bohr's interpretation seems to imply that something is propagating between the locations of the two particles instantaneously. But according to the theory of Special Relativity nothing can propagate faster than light! We will come back to this point in Section 3 and 7 of Chapter 7.

Thirty-four years after the event, on the occasion of the centennial celebration of Einstein's birth, Nathan Rosen, one of the three authors of the EPR argument, wrote a paper analyzing the original EPR paper, as well as Bohr's response. He concluded his appraisal of Bohr's position as follows: "What it seems to amount to is that [according to Bohr] the description of reality by quantum mechanics is complete because reality is whatever quantum mechanics is capable of describing."[9]

7. The Right to Be Wrong

Following the publication in 1935 of the EPR paper and Bohr's response, there were no new, dramatic developments in the Bohr-Einstein debate. Most physicists sided with Bohr, and Einstein found that his position was supported by very few.

All existing records show that neither Bohr nor Einstein budged from his position. However, in view of the general acceptance of Bohr's position within the physics community one wonders: Did Einstein ever show any sign of hesitation?

During the celebrations of the centennial of Einstein's birth in 1979, a special symposium took place in Jerusalem, at which John Archibald Wheeler was one of the participants. John Wheeler, who is now one of the grand old masters of theoretical physics, was associated with Princeton University for many years, including the last years of Einstein's life (Einstein lived in Princeton from 1933 until his death in 1955). After the Jerusalem Symposium was over, Wheeler came to the Ben-Gurion University of the Negev, in Beer Sheva, where I was teaching at the time. My wife and I invited him to a party at our home, and as we sat and talked I asked him: Did Einstein ever show a sign of hesitation in relation to his conviction about quantum mechanics?

Wheeler told me, in response, the following story: In the early fifties there appeared a new formulation of quantum mechanics, called "the path integral approach." This formulation is equivalent to standard quantum mechanics in that it yields the same solution to any and all problems. It is an elegant formulation, however, both mathematically and conceptually. Wheeler, all excited, went to Einstein and told him all about it. Einstein listened patiently. When Wheeler was finished and the discussion ended, he asked Einstein whether this new approach changed his views concerning quantum mechanics. "No," Einstein responded, and then added: "Maybe I have earned the right to make my mistakes."

The drama of the Bohr-Einstein debate drew to a close through the same agencies that end many unfoldings: old age and death. Einstein died in 1955, Bohr in 1962.

Each held on to his views to the very end. They both believed, however, that their argument was a philosophical one; explicitly or implicitly they conceded that quantum mechanics itself is consistent with either position.

They could not have been more wrong.

7. "Nature Loves to Hide"

The Bohr-Einstein debate seemed like a clash between philosophical beliefs. Quite unexpectedly John Bell proved in 1964 that this is not the case. Using a modified version of the thought experiment proposed by Einstein, Podolsky, and Rosen, he settled the argument in favor of Bohr: He proved that whether or not quantum mechanics is complete, it violates Einstein's paradigm of local realism. When the suggested experiment was performed it showed that the quantum mechanical prediction holds true: Nature violates local realism.

~

Nature conceals her secrets because she is sublime, not because she is a trickster.

—Albert Einstein

1. The EPR Thought Experiment: A Reformulation

Forty years after the formulation of quantum mechanics John Bell discovered that hidden within it is a negation of the paradigm of local realism. In 1964 he designed an experiment, a variant of the EPR set-up, which proved that if quantum mechanics is right, then Einstein's paradigm is wrong. It follows that the choice between Einstein's and Bohr's interpretations of the quantum theory is not a matter of philosophical preference; Einstein's view was shown to be untenable. The story of Bell's amazing discovery and its implications is the subject of the present chapter.

The first phase of the story seems innocuous: a reformulation of the EPR thought experiment. A reformulation does not sound like much, but it can open the door to new and unexpected developments. When the great mathematician Leonardo Fibonacci introduced the decimal system in Europe in 1202, he merely suggested a reformulation of the way numbers are written down. In doing so, however, he provided Western science with an indispensable tool: Try to imagine current scientific research being conducted with Roman numerals!

The reformulation of the EPR experiment we are about to describe was suggested in 1957 by David Bohm and Yakir Aharonov. In the Bohm-Aharonov version the experiment starts out with one particle (the "parent particle"), a nucleus of an atom. They assume that the *angular momentum*, or *spin*, of the nucleus is

Figure 7.1   The set-up of the Aharonov-Bohm version of the Einstein, Podolsky, and Rosen (EPR) thought experiment. The "parent particle" disintegrates into two "offspring particles," A and B.

zero. This parent particle is radioactive: At a given moment it disintegrates spontaneously into two "offspring particles," A and B, which move to the right and left, respectively (Figure 7.1). Now recall (Chapter 5, Section 3) that in general the different spin components of a particle cannot have simultaneously well-defined values. For example, if the x-component of the spin is well defined, the y- and z-components cannot be well defined. We pointed out, however, that this principle has an exception: If the spin of a particle is zero, all three components are simultaneously well defined. They all have the value zero—precisely zero.

The argument of the reformulated EPR thought experiment now proceeds just like the original: In the original EPR paper the following two questions are tackled: "Is the position of particle B an element of reality?" and "Is the momentum of particle B an element of reality?" Bohm and Aharonov ask different questions. Their questions are: "Is the x-component of the spin of particle B an element of reality?" "Is the y-component of the spin of particle B an element of reality?" and "Is the z-component of the spin of particle B an element of reality?"

They proceed to demonstrate that the answer to all three questions is yes. Their proof is based on EPR's definition of "an element of reality" (Chapter 6, Section 5): "If, without in any way disturbing a system, we can predict with certainty (i.e., with probability equal to unity) the value of a physical quantity, then there exists an element of physical reality corresponding to this physical quantity." Is it possible, they ask, to find the x-component of the spin of particle B without disturbing it? Yes, it is: Measure the x-component of the spin of particle A. Since the x-component of the spin of the parent particle was zero, and each component of the spin is conserved, the x-component of the spin of particle A and the x-component of the spin of particle B must add up to zero. It follows that if we measure the x-component of the spin of particle A we can deduce the x-component of the spin of particle B without disturbing it. Hence the x-component of the spin of particle B is an element of reality.

The same argument applies to the y- and z-components of the spin of particle B. It follows that all three components are elements of reality; that is, if we accept EPR's definition, all three components of the spin of particle B have well-defined values. But according to quantum mechanics it is impossible even for two of the components, let alone all three, to have, simultaneously, well-defined values. This proves that quantum mechanics is incomplete. Bohm and Aharonov's proof, using spin components, is equivalent to the original version of EPR, which uses position and momentum. Its efficacy will become evident when we discuss Bell's discovery.

2. Dr. Bertlemann's Socks

What can we deduce from the EPR thought experiment? Suppose we accept the EPR criterion of "an element of reality" and admit, somewhat sadly, that quantum mechanics is incomplete. Does our admission constitute an earth-shaking blow to our concept of reality? Hardly. If quantum mechanics is incomplete, let the physicists look diligently for the complete theory behind quantum mechanics, while the rest of us patiently await the results of their investigations.

EPR's proof is based on the fact that the results of experiments carried out at different locations can be correlated: If the x-component of the spin of electron A is found to be "up," the x-component of the spin of B is sure to be "down," and vice versa, with the analogous statements being true for the y- and z-components. Is there anything remarkable or surprising about these correlations? Apparently not. John Bell put it this way:

The philosopher in the street, who has not suffered a course in quantum mechanics, is quite unimpressed by Einstein-Podolsky-Rosen correlations. He can point to many examples of similar correlations in everyday life. The case of Bertlemann's socks is often cited. Dr. Bertlemann likes to wear two socks of different colors. Which color he will have on a given foot on a given day is quite unpredictable. But when you see that the first sock is pink you can be already sure that the second sock will not be pink. Observation of the first, and experience of Bertlemann, gives immediate information about the second. But apart from that there is no mystery here. And is not the EPR business just the same?[1]

Since we know that Dr. Bertlemann is in the habit of wearing socks of different colors on his two feet, our ability to predict, once we see a pink sock on his right foot, that the one on the left will not be pink is no surprise. Similarly, since we know that the spin of the parent particle is zero, it is no surprise that, once we measure a given component of the spin of particle A and the result is "up," we can predict that the same component of the spin of particle B will be "down."

The obvious, commonsense explanation of the EPR correlations is Einstein's: Both particles, A and B, have precise, well-defined values of all three components, but quantum mechanics tells us that we cannot know all three simultaneously. John Bell expressed this view succinctly: "*Nature knows, but I don't know.*"[2]

This explanation is in line with the doctrine of "local realism," a doctrine that combines two assumptions, "realism" and "locality." Realism, you may recall, is the belief that nature exists, in itself, unrelated to the presence of consciousness, and that bits of matter have their own properties. Within the framework of realism the EPR sufficient condition makes perfect sense: Any property of a particle (position, momentum, x-component of spin, etc.) that can be found out without in any way disturbing it must be inherent to it. In the context of the EPR set-up, the alternative claim, namely, the assertion that the measurement of particle A has any effect on the properties of B, implies that the influence of the measurement propagates instantly, faster than light! The assumption that nothing propagates faster than light is called "locality."

As we saw in Chapter 2, local realism is a mainstay of Einstein's paradigm. Einstein was a realist who believed in the findings of the theory of Special Relativity. And since according to Special Relativity the speed of light is the maximum possible speed, he believed not only in realism but also in locality.

Einstein's explanation of the EPR correlations is straightforward, yet Bohr adhered to a different explanation. According to Bohr, unless a measurement is carried out, neither particle A nor particle B has any component of spin. If the x-component of the spin of A is measured, the measurement *creates* this x-component of A, and, because the two particles are entangled, the same measurement also creates the x-component of the spin of B. And if the y- rather than the x-component of A is measured, then both A and B will continue not to have well-defined values of the x-component.

Unlike Einstein's explanation, Bohr's view violates locality; it seems to imply that when a measurement is carried out on particle A, it affects particle B *instantaneously*, even if particle B is light-years away! Yet this view became, early on, the mainstream view among physicists. Why was Bohr's view given credence, let alone dominance, when Einstein's view, which is the embodiment of common sense, explains the EPR correlations quite well?

In 1932 the mathematician John von Neumann proved that theories that are in line with Einstein's explanation, called "hidden variables theories," are incompatible with quantum mechanics. Was this the reason for the dominance of Bohr's view? No, it wasn't: When David Bohm showed in 1952 that von Neumann was wrong by constructing a hidden variables theory that was compatible with quantum mechanics (that is, constructing a theory that von Neumann "proved" cannot be constructed), his paper caused hardly a stir. It seems to me that many physicists, myself included, preferred Bohr's view because we *felt* that quantum mechanics is complete and contains a radical message about the nature of reality. Although scientific claims are always presented in rational terms, they are often based on feelings and intuitions rather than logical arguments! At any rate, by the early sixties the issue caught the attention of a young Irish physicist, John S. Bell, who decided to clear the murky waters. He believed in Einstein's explanation, set out to show that the objections to it are all wrong, and ended up proving that it is Einstein's explanation which is untenable and has to be given up.

3. Bell's Modifications of the EPR Thought Experiment.

For decades Niels Bohr exerted his enormous intellectual power in an effort to prove to Einstein that quantum mechanics calls for a radical change in his (Einstein's) paradigm. At the end he had to admit, at least implicitly, that Einstein's position was logically defensible: It is possible to maintain one's belief in Einstein's paradigm if one takes the not unreasonable position that quantum mechanics is incomplete. And yet, two years after Bohr's death, John Bell managed to discover what eluded Bohr in spite of decades of trying: an ironclad, mathematical proof that Einstein's paradigm is indeed incompatible with quantum mechanics. It does not matter whether quantum mechanics is complete or incomplete; if one accepts that quantum mechanics is *correct* (and Einstein accepted that), a profound change in Einstein's paradigm is unavoidable.

The significance of Bell's work was a little slow in reaching his fellow physicists. This is probably due to the fact that his 1964 paper was written using a complex statistical formalism, and its claim referred to the inadmissibility of hidden variables theories, theories that were of little interest to most physicists. Hidden variables theories are based on EPR's assertion that the quantum theory is incomplete. They claim to complete it by introducing additional quantities, or variables, which, unlike the standard variables of the quantum theory, are not observable but are "hidden." Most physicists paid little attention to the hidden variable theories that were proposed; they seemed a bit contrived. None of them seemed to be convincing enough to be accepted as the correct "completion" of quantum mechanics.

Within a few years, however, the message of Bell's paper was clarified, and the proof of its main point turned out to be amazingly simple. Both the message and the proof are beautifully explained by Bernard d'Espagnat in a 1979 article in *Scientific American*.[3]

To make his point, Bell introduced the following three modifications into the Bohm-Aharonov version of the EPR experiment:

First, instead of considering the components of the spins of the two particles, A and B, along the x-, y-, and z-axes, that is, along three axes which are perpendicular to each other, he considered the components of these spins along any three axes, a, b, and c, the angles among which are not mutually perpendicular (see Figure 7.2).

Second, he introduced *random decision-makers* in front of the two apparatuses that measure the spin components of particles A and B (see Figure 7.3). Random decision-makers are devices that decide randomly which spin component, a, b, or c, will be measured. These devices are constructed and placed in such a way that the decision made by the device near A is made and communicated to the A-apparatus a very short time before the measurement on particle A takes place. The time is so short that if a light signal leaves this device and starts traveling toward the measuring apparatus of particle B, both the A-measurement and the B-measure-

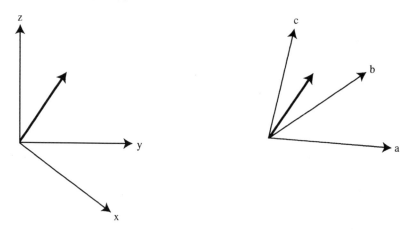

Figure 7.2   The components of a vector are usually taken along three mutually perpendicular axes, x, y, and z. They can be taken, however, along other sets of axes, such as a, b, and c, which are not at right angles to each other.

Figure 7.3   The EPR set-up, as it was modified by Bell. Some component of the spin of particle A and some component of the spin of particle B are measured simultaneously. The decisions as to which component will be measured are made by "random decision-makers." Such pairs of measurements are repeated many times, and the results are analyzed statistically.

ment will have been completed before that light signal reaches the B-apparatus. The construction and placement of the random decision-maker near B is subject to a similar stipulation.

The purpose of this stipulation is to rule out the possibility that the random decision made near particle A would affect the measurement of particle B, and vice versa. If one believes in locality (i.e., if one believes that nothing propagates faster than light), then this stipulation guarantees that the measurements at A and at B do not affect each other: Since the fastest signal that can travel between A and B travels at the speed of light, the stipulation ensures that the measurement of B will be complete before a signal from A could possibly reach B, and the measurement of A will be complete before a signal can reach it from B.

And third, he envisioned doing the experiment not once, but numerous times.

What was the purpose of these modifications? Bell found that by introducing them, and analyzing the results of the many measurements statistically, he could come up with certain relations that *must be satisfied if Einstein's paradigm holds, but are not satisfied when calculated on the basis of quantum mechanics.* He thus proved that *quantum mechanics is incompatible with Einstein's paradigm of local realism.* These relations are called "Bell's inequalities," and the fact that they are not satisfied when calculated on the basis of quantum mechanics is known as "Bell's theorem" and also as "the violation of Bell's inequalities."

4. The Case of the Contrary Couples

What is "an inequality"? An inequality is a statement of a relation in which the quantities involved are unequal to each other. For example, the statement "The number five is greater than the number three" is an inequality. The inequalities that Bell discovered are slightly more complex. They say, "One quantity is less than or equal to (but never greater than) the sum of two other quantities." Let us start out with an example of such an inequality.

I don't like to be short of quarters and dimes, so I have developed a habit of keeping some quarters in a "quarters box," and some dimes in a "dimes box." Whenever I have a few extra quarters or dimes I drop them in the appropriate box. Yesterday morning I noticed that my stash of quarters and dimes was low: There were two quarters in the quarters box, and two dimes in the dimes box. I don't remember whether or not I added any coins to either box since yesterday morning, but I do remember that I did not take any coins out. What I now know about the contents of the boxes can be stated as an inequality: "The number four is smaller than or equal to the sum of the numbers of coins in my two boxes."

Bell introduced his three modifications of the EPR set-up because the modified set-up enabled him to formulate inequalities that provide a test of the assumption of local realism. As we will see in the next section, Bell applied his test to spin components, but the test itself is of general applicability: Bell's inequalities can be used to check on the local realism of a great variety of systems.

We will introduce Bell's inequalities through a story about people, rather than particles. As you read the story, remember that it is fiction, and that the couples (husbands and wives) who feature in it are stand-ins for pairs of particles. In other words, the moral of the story has nothing to do with people; I chose to introduce Bell's idea through a psychological rather than physical set-up merely because, for most of us, people are more interesting than electrons. So, here we go:

A long time ago, before my good friends Julie and Peter, with whom you became acquainted back in Section 6 of Chapter 2, started their astronaut training, they were seniors in college, living in New York and studying psychology. One sunny afternoon, as they were strolling down Broadway, they happened to be walking behind an elderly couple and could not help overhearing a strange conversation:

"You know, I think the president is doing a terrible job," said the man.

"The president? I think he is doing a great job," responded the woman.

"No. If it wasn't for Congress, the country would be finished!"

"Hardly. I think they are the problem!"

"Well, you have to admit that the Speaker is doing us some good."

"I really don't think he is. He's as bad as anybody else in the Congress."

"Well, at any rate, I support his proposed changes in the income tax law."

"I feel that his proposal is completely inappropriate."

The conversation continued in this vein, a string of affirmations and negations, for a long time. Strangely, the man and the woman were not arguing with each other. They were just cruising from one subject to the next, always stating contrary opinions. They were speaking in low, tired voices, as if they had been through this kind of conversation many times, were not interested in it anymore, and yet felt compelled to go through the motions.

The next day Julie and Peter ran into their favorite psychology professor, Clarissa Gill. When they told her about the strange conversation they had overheard, Clarissa sighed.

"Yes," she said, "these poor people are suffering from SCC, the Syndrome of the Contrary Couples. There are hundreds if not thousands like them; I myself have assembled a sizable directory. Their sickness consists in having a strong and adversarial psychic relationship that compels them to contradict each other at every opportunity. When you ask them about *facts*, they would agree. But ask for their *opinion* about anything, and they are sure to give you contrary responses. When the syndrome first sets in," Clarissa added, "they fight a lot. After a while they become accustomed to their situation, but they simply can't help having these conversations in which all they do is contradict each other."

"But why do they always have to *state* these contrary opinions?" Julie wondered. "Are they angry at each other?"

"They may be," Clarissa responded, "but this does not seem to be the crux of the problem. In fact, it was experimentally established that they give contrary answers to questions of opinion even when they are interrogated separately, when the husband has no idea about his wife's response, and vice versa."

"Wait a minute," Peter said, "are you telling me that if I meet the husband downtown at the same time that Julie meets the wife in the Bronx, and we each ask the husband and wife, 'Do you approve of the president's performance?' you can be quite sure that if the husband says yes, the wife would say no?"

"Absolutely. And if Julie hears the wife say yes, she can be sure the husband said no to you."

"Wow, that's so weird! How does it happen?"

"I have no idea, but I am trying to find out. That's why I assembled this directory of contrary couples. Using this directory, I intend to carry out some experiments."

Julie and Peter responded almost at the same instant. "Can I help?"

Clarissa's face broke into a broad smile. "That would be wonderful. I could really use your help. Can you come to my office tomorrow at ten o'clock?"

The next morning Professor Gill explained to Julie and Peter the experiments she had in mind. "Early this year," she told them, "I discovered something: If we pick three questions and a fairly large sample of contrary couples, and ask them these questions, the husbands' and wives' responses have to follow a certain pattern."

"Let me get this straight," Peter interrupted. "Suppose I choose these three questions: (a) Should the size of the federal government be reduced? (b) Is the secretary of state doing a good job? and (c) Should the minimum wage be raised? Are these questions okay?"

"Yes, they are. As I told you, any question of opinion would do."

"Okay, so then suppose we choose a large number of contrary couples and interview them a couple at a time. For instance, tomorrow at 11:00 sharp Julie will speak to the wife here in your office, and I will meet her husband in my apartment in the Bronx. Both of us will get yes or no answers to my three questions. At 11:05 we interview another couple, at 11:10 another, then another. . . . And after all that, the data we collect should follow this pattern that you discovered?"

"That's right."

"And what is the pattern?"

"I will explain it to you after you complete the interviews."

So Peter and Julie went to work. They started interviewing contrary couples, one after the other, and, unexpectedly, they ran into trouble—big trouble. When the three of them met again, a week later, Julie could not hide her frustration. "Professor Gill," she said, "you won't believe what happened. In my very first interview, the wife entered, I asked her question (a), and she responded, Yes. But when I asked her question (b), she said, 'One question is enough,' and walked out. I tried to stop her, but she wouldn't listen. The second woman was even worse. She told me, as she walked in, 'I am going to answer only one question.' And so it went. I couldn't get any of them to answer more than one question, and Peter had the same experience with the husbands. What did we do wrong?"

Clarissa sighed. "Don't take it personally," she said, "it's not your fault. Just this morning I was reading an article about the Contrary Couples Syndrome and found out that one of the symptoms is an absolute refusal to answer more than one question. These people are really sick."

"So what are we going to do?" Julie was close to tears.

Clarissa smiled. "Just before you came I found a way around this problem," she said. "First, even if a wife would answer only one of your questions, you can find out about her opinion regarding two questions, say, questions (a) and (b), by asking her question (a) and asking her husband question (b). His answer would reveal her predisposition because it has to be contrary to his answer."

"I see. So if the husband says yes in response to question (b), we know that she would have said no, and vice versa. But how do we get the third question in?"

"We can't find each wife's predisposition regarding all three questions, but that doesn't mean we can't still check whether the pattern I discovered holds true. To do this, both of you need to bring a die to the interviews."

"A die?!"

"Yes. Dice can serve as random decision-makers. Since you are allowed only one question, you need a device that will decide for you, randomly, which question to ask. Before you begin an interview, you roll the die. If 1 or 2 comes up, you ask question (a); if 3 or 4 comes up, you ask question (b); and if 5 or 6 comes up, you ask question (c). And Peter will do the same, simultaneously."

"What good will that do?"

"After you and Peter repeat the procedure many times, we will be able to do some counting. We will be able to count how many times the wife responds yes to question (a) and the husband simultaneously responds no to question (b), how many times the wife responds no to question (b) while the husband responds yes to question (c), and so on and so forth."

"And what's the significance of that data?"

"Well, hopefully, this will provide experimental proof of the pattern I discovered."

"Hey, wait a sec, I just remembered something. I once overheard someone speaking about something called 'Gill's inequality.' Is this the pattern you discovered?"

"Yes, Peter," Clarissa said, "the pattern I was talking about is an inequality. This inequality involves the results of three counts. It goes like this: Suppose you and Julie conduct lots and lots of your simultaneous interviews. When you are finished, you do some counting. First you count the number of times that, after casting her die, Julie asked the wife question (a) and she answered yes, and you, Peter, following your die, asked her husband question (b) and he responded no. Is this clear?"

"Yes," Peter responded. "You want us to count the number of correlations between wives' positive responses to question (a) and their husbands' simultaneous negative responses to question (b)."

"That's right." Peter was pleased with himself. He had used the term "correlation" correctly! "Second," Clarissa continued, "you count the number of correlations between wives' yes in response to question (a) and husbands' no in response to question (c). And, third, you count the number of wives' yes in responses to question (b) and simultaneous husbands' yes responses to question (c). Now you compare the three counts. My inequality says that the number you get by adding the second and third counts together is either greater than or equal to the first count, never smaller."

"And just how do you go about proving something like this?"

"Well, the proof takes some doing. I will let you work it out yourselves. As you try to prove it, remember the two assumptions it rests on: The first is locality: Nothing travels faster than light. The second is the existence of predispositions: I assume that, although both husbands and wives refuse to answer more than one question, they have clear predispositions for a yes or a no response in relation to all three questions."

After some hard thinking, Julie and Peter managed to prove Clarissa Gill's inequality to their satisfaction. An account of the proof is given in Appendix 2. Here we continue with our story.

Julie and Peter then conducted lots and lots of interviews, exhausting Clarissa's directory of contrary couples. When they collected the data and did the three counts, they were surprised to discover that Gill's inequality was violated: The first count turned out to be more than the sum of the other two! They appeared in Clarissa's office looking puzzled and confused.

Clarissa admitted that she too was surprised. All three of them just sat there, thinking. Eventually Clarissa said: "Well, one of the two assumptions I used in proving my inequality must be wrong. It is possible, for example, that the answers we receive are not due to predispositions but are the result of a whim of the moment."

"But in this case, how can the answers always be contrary?" Julie objected. "Remember, the wife and husband responded simultaneously at separate locations. There was no communication between them."

"Hmm, I know. This is truly puzzling!"

This is the end of my story. Its psychological context is merely an analogy: The contrary couples represent pairs of particles, the A and B particles of EPR. Admittedly, the human characteristics I chose to invoke are a bit forced (the assumption of perfect contrariness, especially in the absence of communication, is hardly realistic), but, as we will see in the next section, the analogy with the EPR situation is almost perfect.

5. Bell's Inequalities

Bell's inequalities refer to pairs of particles in the EPR set-up, rather than husbands and wives. The logic that leads to the inequalities, however, is the same in both cases: The three questions we discussed in the previous section correspond to the three spin directions a, b, and c. The yes and no responses to the questions correspond to the result of measurements of any spin component. A yes corresponds to the result of the measurement being "up," a no to the result being "down." The refusal to answer more than one question corresponds to the impossibility of measuring more than one of the three spin components. The assumption of predispositions corresponds to the assumption that each particle has, as "an element of reality," a specific component, "up" or "down," in all three directions. And, finally, the fact that Gill's inequality is violated corresponds to the violation of Bell's inequalities (see below), a violation that leads to the conclusion that in the EPR set-up the assumption of local realism is invalid.

To formulate Bell's inequalities, all we need to do is "translate" Clarissa Gill's three correlation counts from the language of contrary couples to the language of spins.

Suppose we conduct the EPR experiment with Bell's modifications and repeat it many, many times. Then we collect the results and do some counting. We start out by counting the number of correlations between particle A having an a component "up" and particle B having a b component "down." That is, we count the number of times that one "random decision-maker" instructed the A-apparatus to measure the spin component of particle A in the direction a, and when the measurement was carried out the result was "up," and simultaneously, the other random decision-maker instructed the B-apparatus to measure the spin component of particle B in the direction b, and when the measurement was carried out the result was "down." Then we do two more counts. We count the number of correlations between particle A having an a component "up" and particle B having a c component "up." And, finally, we count the number of correlations between particle A having a b component "up" and particle B having a c component "down." Now we are ready to state one of Bell's inequalities.

A Bell's inequality: *The number we get by adding up the results of the second and third counts is either greater than or equal to the first count, never smaller.*

If you will take the time to read Appendix 2, you will verify for yourself that there is nothing mysterious about Bell's inequalities. Once the assumption of local realism is accepted, the proof amounts to nothing more than a sophisticated exercise in counting. The amazing fact is that, for some choices of the directions of the axes a, b, and c, such as the choice we see in Figure 7.2, the calculations of the correlations according to quantum mechanics yield counts that *violate* Bell's inequalities. For example, for the three directions of Figure 7.2 the quantum mechanical calculation shows that the sum of the second and third counts should be *considerably smaller* than the first.

How can that be? Well, remember the contrary couples. The violation of Gill's inequality means that the assumption of predispositions or the assumption of locality (which says that instantaneous communication between husband and wife is impossible) is invalid, or perhaps that both are. Analogously, Bell's inequalities are based on the assumptions of local realism: The spin components are "elements of reality" in the sense of EPR's definition, and no influence can propagate faster than light. The violation of Bell's inequality means that at least one of these assumptions is wrong. Which assumption has to be given up? What kind of reality do we end up with, once we do give up one (or both) of these assumptions?

If you followed the main line of development in this chapter (whether or not you followed all the details—that does not really matter) you probably feel at this point as I felt when I first grasped the significance of Bell's correlations, and as many other physicists feel—rather uneasy. It feels as if the ground is slipping from under our feet, and we are not sure where we will land. *This puzzlement and this feeling are characteristic of the first stage of a paradigm shift.* Our view of reality will have to change. The commonsense view of reality leads directly to Bell's inequalities—there is no room to maneuver—and yet quantum mechanics, which has been verified by thousands of accurate experiments, says that they can be violated!

What's going on? We will keep probing for an answer in the coming chapters. For the time being let us raise a different question: We have a confrontation between Bell's inequalities and quantum mechanics. Is it possible to carry this confrontation into the experimental domain, that is, perform an experiment with pairs of particles, A and B, measure the correlations, and see whether the inequalities are satisfied or violated?

6. The Experimental Situation

So far we have treated Bell's version of the EPR set-up as a thought experiment. As we have seen, Bell managed to prove rigorously, on the basis of this treatment, that Einstein's paradigm of local realism is incompatible with quantum mechanics. The discovery of such a proof is an impressive achievement. Bell's contribution was, however, a double whammy. Not only did it lead to the important result just mentioned; it turns out that, unlike the experiments that Bohr and Einstein argued about, the experiment he proposed was not just a thought experiment—it was almost doable.

The significance of actually carrying out the experiment is this: Bell demonstrated that for certain choices of the axes a, b, and c the inequalities he derived on the basis of Einstein's paradigm are violated by quantum mechanics. This shows that the two, Einstein's paradigm and quantum mechanics, are incompatible. But which is right? Which does nature follow, Einstein's paradigm or quantum mechanics? The actual performance of the experiment should provide an answer to this question.

I wrote that the experiment was "almost" doable because, as J. Clauser, M. Horne, R. A. Holt, and A. Shimony have shown, the experiment that was doable was not the one we described, involving nuclei of atoms, but an equivalent one involving photons. The first such experiment was carried out by J. Clauser and S. Freedman in 1972. More experiments were conducted during the seventies. These experiments tended to confirm quantum mechanics, but their level of accuracy was not high enough to render their findings conclusive. The breakthrough came in 1981, when A. Aspect, P. Grangier, and G. Roger, working at the University of Paris, performed an experiment that was much more accurate than the preceding ones. They summarized their results as follows: "Our results, in excellent agreement with the quantum mechanical predictions, strongly violate the generalized Bell's inequalities, and rule out the whole class of realistic local theories."[4]

Einstein's paradigm of local realism is the world-view that we live by. Aspect and his co-workers have demonstrated experimentally that this is not the paradigm that nature adheres to. What, then, is the new world-view? In the next chapter we will start facing this question in earnest. The remainder of the present chapter will be devoted to further clarifications regarding the EPR experiment and Bell's inequalities as well as to recent developments related to them.

7. Influences and Signals

According to the theory of Special Relativity, the speed of light is the highest speed at which communications can be carried out. For example, those of us who watched Neil Armstrong take his first few steps on the moon saw his every movement at least a second and a quarter after it took place. The distance between the earth and moon is such that it takes light about a second and a quarter to cover it. There was no way to find out what was going on with Armstrong right then. If the theory of Special Relativity is right, this delay of a second and a quarter can never be reduced.

It seems, however, that, according to EPR this delay *can* be reduced—all the way down to zero—by using the EPR set-up. If Bohr is right and Eisntein is wrong, a measurement performed on particle A determines, *instantly*, the corresponding quantity of particle B. For example, if the *x*-component of the spin of particle A is measured and found to be "up," this very measurement *makes* the *x*-component of particle B be "down." Some influence must be propagating, then, between A and B instantly. Why not use this influence to transmit information faster than light? It seems as if something is propagating between A and B instantly. Why not use this something to let us know what an astronaut on the moon, or way out in space, is doing *right now*?

My friend Peter, the psychology-major-turned-astronaut, decided to try. A few days before he and Julie were scheduled to embark on two separate missions, he came up with an idea: "Let's use the EPR set-up to communicate instantly," he suggested to Julie. "The way our communications are set up now, if we were, say, a full light-hour apart the transmission of a message would involve a whole hour's delay—that's just too long."

"Yes," Julie responded, "a light-hour is such a great distance that it takes even light a whole hour to cover it. And since nothing travels faster than light, there is no way for my message to reach you in less than an hour, or so it would seem. But I too would like us to be in closer touch. Do you have a way to get around the speed-of-light barrier?"

"I think I might; check this out. Suppose we load our spaceships with the two apparatuses of the EPR experiment. I carry the A-apparatus, you carry the B-apparatus. As you know, our spaceships are scheduled to travel away from earth and away from each other. Suppose we arrange for the EPR experiment to be carried out on earth many times while we are traveling. We will also adjust the speeds of our spaceships so that the time of arrival of the A particle at my apparatus will coincide with the time of arrival of a B particle at yours. When these particles arrive, both of us will measure the same component of the spin of our particles. The results of our measurements will be anti-correlated: Whenever I will get the result 'up,' I know that you will get 'down,' and vice versa. That's a kind of communication already. That in itself will make me feel connected to you."

"That's very nice, but you don't need such a complicated arrangement for that. We can just decide that at a certain time that we agree on, we will look at each other's pictures and feel connected. What I'm after is real communication, my message reaching you instantly!"

"I agree," Peter responded, "and here is how I'm going to send you a message. Suppose we follow my suggestion, use the EPR set-up and keep measuring the same component, say, the x-component, time after time; each measurement will give us a transmission of one bit of information. Suppose we use a binary code—change our verbal message to a sequence of o's and 1's. Is that okay?"

"So far, so good."

"Now, whenever the result of your measurement is 'up,' it means that I sent you a o. Whenever the result is 'down,' it means that I sent you a 1. You keep taking measurements and get a sequence of o's and 1's, from which you can decipher my message!"

Julie was fascinated, but an uneasy feeling at the pit of her stomach warned her that something didn't add up. She concentrated, trying to visualize herself deciphering Peter's message. Then she visualized Peter sending those messages, and as soon as she did that, she understood.

"No, Peter," she said, "it won't work."

"But why?"

"Because," she responded, "when you, at your end, are using your measuring apparatus, you can't *determine* the result of the measurement. To send me a o, you have to make the result of your measurement be 'down,' so that my result will be 'up.' But you can't do that. All you can do is measure. Sometimes the result is 'up,' and sometimes it's 'down.' You have no control over which one it will be. You can't send me messages!"

Peter was crestfallen. "I guess you're right." he finally said. "Both of us are merely passive readers of these results. When you wish to communicate, you have to be active, deciding what to transmit. We can't do that with the EPR set-up, so we can't communicate faster than light."

"Don't be sad, Peter. We can't use the EPR experiment to communicate, but next Wednesday at noon, when we will be a light-hour away from each other, I'll call you on the intercom."

"Thank you, Julie. I won't forget. I'll be waiting for your message at one o'clock sharp."

The story illustrates the fact that, on the one hand, some influence is apparently propagating between A and B, and, on the other, this influence cannot be used to transmit information, deliver messages, or send signals. This led Henry Stapp to make a distinction between "influences," which can move faster than light, and "signals," which cannot. The distinction has to do with control. As Peter discovered in his valiant attempt, that which (apparently) propagates between A and B is uncontrollable. It is an influence. Signals, however, are controllable. Peter decides what message he wishes to send Julie; he controls the message. Signals are subject to the limitation of locality, and influences are not.

The distinction between signals and influences is a subtle one. It feels as if, in allowing influences the freedom to travel faster than light and denying it to signals, nature is walking a very fine line. On second thought, nature, most likely, is merely hiding; it knows very well what it is doing. It is our understanding that walks a very fine line. For example, we have just referred to the faster-than-light transmission of influences as "apparent." Is it merely apparent, or is something really transmitted faster than light? The fact that in the case of influences no pulse of energy between A and B was ever detected indicates, but does not prove, that we are not dealing with a real transmission. We will come back to this equation in Section 9 of Chapter 16.

8. Recent Developments

As we have seen, statistics features prominently in Bell's proof. Bell's inequalities are relationships among statistical correlations. Certainly, there is nothing wrong with that. When statistics is used correctly, it is a reliable tool of analysis. Nevertheless, some physicists raised the question as to whether the incompatibility between Einstein's paradigm and quantum mechanics can be exhibited *in one measurement*, rather than through statistical methods, which require many measurements.

In 1988 D. M. Greenberger, M. A. Horne, and A. Zeilinger (GHZ is the shorthand for their names)[5] came up with a thought experiment that does just that. The price paid for eliminating statistics is the need to measure the correlations among three particles, rather than two. The three particles are assumed by GHZ to originate from one parent particle, in such a way that the angular momentum is conserved—just as in Bell's experiment. It then turns out that if any of their spin components are measured (choose any component you like—but keep the same component in all three measurements), then Einstein's paradigm and quantum mechanics lead to different predictions regarding the correlation of the three measurements.

The GHZ experiment is elegant, but—unlike Bell's—it is at present a thought experiment. No suggestion as to how to carry it out as an actual experiment has yet been put forth.

Three years after the GHZ proposal Lucien Hardy proposed an experiment involving only two photons that, in an ideal situation, can test Bell's theorem without the need for statistics.[6] The experiment needs to be carried out only once. This requires, however, the use of ideal detectors, i.e., detectors of extremely high efficiency. The same experiment can be done using available, as opposed to ideal, detectors, but then it would be necessary to repeat it many times and use statistics.

Perhaps the most interesting recent development is the discovery of a way to use the EPR experimental set-up, also known as "the EPR entanglement" of the two particles involved, to transport quantum states from one location to another. The theoretical possibility of such "quantum teleportation" was discovered in 1993 by C. H. Bennett and his collaborators. This possibility was realized experimentally, for photons, five years later by two experimental groups.[7]

A quantum teleportation experiment begins with two EPR-entangled photons. Photon A travels to Laboratory 1 and photon B to Laboratory 2, which can be located far away from Laboratory 1. The aim of the experiment is to "teleport" the quantum state of a third photon C, which exists in Laboratory 1, to Laboratory 2— without, however, physically bringing photon C to Laboratory 2.

It turns out that this teleportation can be achieved by entangling photon C with photon B. This new EPR entanglement changes the state of the faraway photon A. Now, if certain measurements are carried out on photons A and C in Laboratory 1, and the results are conveyed to an experimenter in Laboratory 2, he or she can carry out an appropriate adjustment of the state of photon B. The result of the adjustment is making the quantum state of photon B identical to the original quantum state of photon C.

In such a quantum teleportation experiment the entanglement of photons A and B creates a channel for the transmission of information between laboratories that are far apart. This transmission seems to be instantaneous, but that can't be right: Such an instantaneous transmission of information would violate Special Relativity. A closer look at the experiment reveals, however, that the transmission of information is not instantaneous. The teleportation cannot be achieved without the transmission of some information from an experimenter in Laboratory 1 to an experimenter in Laboratory 2 using ordinary means of communication, such as a phone or a fax.

9. In a Minor Key

It is easy to imagine how delighted Bohr would have been had he lived to learn of Bell's discovery. It is not so easy to imagine Einstein's response. I wonder about it from time to time, and I don't have a clue.

Einstein died in 1955, but one of the co-authors of the EPR paper, Nathan Rosen, lived to a ripe old age and died in 1995. I met him for the first time in 1971 and asked him what his view on the EPR paper was. He said that he still considered it correct but did not feel as strongly about the issue as he used to. In a paper written a few years later he was more explicit. He still expressed the belief that the incomplete quantum theory will be completed by a more fundamental one. He acknowledged, however, that, in view of Bell's discovery, this complete theory is unlikely to be of the "hidden variables" variety, and that the completion is unlikely to be easy. He concluded his paper in a minor key:

What are the prospects of finding a satisfactory theory that will give a complete description of reality? One must not be overly optimistic. It appears that such a theory will not be obtained by some simple modification of quantum mechanics, such as the addition of hidden variables. If someday quantum mechanics is replaced by another theory, this is likely to involve revolutionary changes in concepts and principles—perhaps even changes in our concepts of space and time. In that case it may even turn out that the question posed by the [EPR] paper—is the description of physical reality complete?—no longer has a meaning, or that it has to be given a different interpretation. The consequences of a revolution in physics are hard to foresee.[8]

PART TWO
From a Universe of Objects to a Universe of Experiences

The violations of Bell's inequalities prove that nature does not adhere to the paradigm of local realism. This presents us with a quandary: We have to let go of local realism, but do we have a viable alternative?

The journey toward an alternative world-view begins with a close look at the assumption of realism. Two great thinkers of the twentieth century have pointed out that realism is not a fundamental truth about the nature of reality. Alfred North Whitehead demonstrated that concrete facts are experiences rather than objects; and Erwin Schrödinger pointed out that a scientific inquiry always starts by removing the "Subject of Cognizance" from its domain and treating its subject matter as a lifeless object. He called this procedure "the principle of objectivation."

Whitehead's and Schrödinger's analyses support the conclusion arrived at in Part I: Local realism has to go. But what will replace it? Since quantum mechanics itself is based on the principle of objectivation, it cannot provide the alternative. Quantum mechanics, however, contains a deep mystery. The mystery, which is known as "the collapse of quantum states," is the process whereby the potential becomes actual. An analysis of the process of collapse indicates that the principle of objectivation has to be transcended, and, in addition, it provides broad hints about the alternative paradigm. Similarities between the process of collapse and the ideas of Plato lead to the following questions: Is it possible that seemingly inanimate objects are alive? Was Plato right in believing that the whole universe is alive? How can we answer such questions?

Part II culminates in a presentation of Whitehead's magnificent "process philosophy" and an analysis of its astonishing correspondence with the findings of quantum physics. Process philosophy is a rigorous and inspiring metaphysical system that claims that the universe and its constituents are alive; the "atoms of reality" are "throbs of experience." Whitehead's system resolves the quandary presented in Part I: Realism is merely an abstraction. We live not in a universe of objects but in a universe of experiences.

8. The Elusive Obvious

John Bell showed that local realism has to be given up, and a close look at Bell's theorem shows that locality cannot be saved. But should we try to keep realism as an aspect of the new, emergent paradigm? Alfred North Whitehead's perceptive analysis of "the fallacy of misplaced concreteness" demonstrated that realism is not a fundamental characteristic of reality. It is not objects that are concrete, but experiences.

~

This fallacy [the fallacy of misplaced concreteness] is the occasion of great confusion in philosophy. It is not necessary for the intellect to fall into this trap, though in this example there has been a very general tendency to do so.

—Alfred North Whitehead

1. The Schizophrenic Physicists

If you read the preceding chapters in a critical frame of mind, you may have detected a strange inconsistency in relation to the Einstein-Bohr controversy. On the one hand, we mentioned the fact that most physicists sided with Bohr. On the other hand, we said that Bell's discovery, which vindicated Bohr's position, caused not only excitement among physicists but also a certain level of puzzlement and discomfort. This is strange. If a theorem or an experiment confirms one's view, one should feel satisfaction rather than discomfort!

This question is related to another question: Why, in spite of Einstein's enormous prestige, as well as the logical consistency of his position, did most physicists side with Bohr?

I would like to bring my own experience to bear on this question. When I studied quantum mechanics as an undergraduate, I, too, sided with Bohr. I felt that Bohr's approach endowed quantum mechanics with a certain profundity, even grandeur. Studying quantum mechanics with Bohr's view in mind, I had the feeling that the theory provided a key to the mystery of the universe—that studying quantum mechanics was tantamount to turning the key, and, if one studied deeply enough, the door would be opened. And, above all, Bohr's interpretation simply *felt* right. By contrast, the acceptance of Einstein's view meant that quantum mechanics is a rather arbitrary mathematical structure, that its different compo-

89

nents are merely elements that just happen to work together, and that there is nothing deep or significant about it. From this perspective the theory is a mere "cover story" for the real thing—the "complete" quantum theory, about which nobody has a clue.

This feeling stayed with me over the years; and through conversations with many physicists—from Nobel Prize laureates to undergraduate students—I became convinced that the preference for Bohr's view among most physicists is due to similar intuitive reasons. But then, it is also true that the vast majority of physicists have not thought about the foundations of the quantum theory very deeply. What we call "quantum mechanics" is really two systems of thought: first, equations and procedures that tell us how to calculate quantities that we wish to know, and second, a conceptual framework regarding the subatomic, atomic, and molecular domain—a framework that, at least in certain respects, can be extended all the way to everyday life, and even to astronomical scales. As far as the first system of thought is concerned, there is no controversy. If Einstein and Bohr wished to calculate the same quantity, they would have used the same equations and the same procedures. These equations and procedures constitute the bulk, if not the entirety, of most of the quantum mechanics courses taught at colleges and universities. Most physicists end up giving *some* thought to the conceptual aspect, on the basis of which they prefer Bohr's position. But since this conceptual aspect plays no part in the day-to-day application of the quantum theory to problem solving, these conceptual questions are usually left on the back burner and are forgotten there—until something begins to burn.

The alarm of an unexpected smell is precisely what the excitement regarding Bell's inequalities is all about. Einstein's paradigm is, in its general terms, everybody's commonsense paradigm, the paradigm that we live by. In its operational terms, it is also the paradigm of the present-day physicist. Bohr's approach is in stark contradiction to it. So there is a contradiction between the physicist's "operational paradigm" and Bohr's approach to quantum mechanics that he or she claims to subscribe to. *But unless one thinks very deeply about quantum mechanics, one may overlook the contradiction with impunity.* In this respect (as in most respects) physicists are like ordinary people. If they can't resolve a contradiction, and the contradiction is not pressing, they just disregard it and give their attention to those aspects of the theory (or theories) that are pleasantly consistent.

This state of affairs is the reason for the title of the present section. Most physicists live in a kind of "schizophrenia" (which literally means "split mind") in relation to the quantum theory. They live, day to day and moment to moment, relying on Einstein's paradigm, but they also adhere to Bohr's interpretation of quantum mechanics. These two contradictory tenets of belief seem to exist in their minds side by side, separated by a watertight partition.

Now we are in position to appreciate both the excitement and the discomfort that Bell's discovery has triggered. *The watertight partition has been punctured.* Adhering to Bohr's view no longer means just the pleasure of engaging in a fascinating philosophical conversation, third martini in hand. It now means, rather, facing the fact that apparent faster-than-light propagation of influences (see Chapter 7, Section 7) can be observed in the laboratory down the hall!

One has no choice but to give up Einstein's paradigm. In this sense our position is analogous to that of the people of Magellan's generation: Once Magellan's ships completed their journey around the earth, the flat-earth paradigm became untenable. We are now at the point where Einstein's paradigm has become untenable, but what will it be replaced by? As far as the atomic and subatomic domains are concerned, Bohr's approach works. But how should it be extended to become a paradigm of reality, a comprehensive world-view?

2. Locality and Realism

Let us go back to Julie and Peter. They have completed their missions in space and are back on earth, spending a well-earned vacation together. Taking a leisurely walk in late afternoon, they are discussing Bell's inequalities and Aspect's experiment.

"You know, I've been putting a lot of thought lately into what Bell's discovery means. Maybe we could try to help each other understand it," Peter began. "Bell's inequalities are based on the assumption of local realism, which is really two assumptions, locality and realism, right?"

"Just to make sure we're on the same page here—what do you mean by these terms?"

"Let me see. 'Locality' means that no signal, no communication, no influence—*nothing* can travel from one place to another faster than light. 'Realism,' as I understand it, is the assumption that the objects of the physical world have an existence of their own, independent of acts of consciousness, and that they have attributes that are also their own, like location and shape."

"So, when you speak about realism, Peter, you're saying that that building we see over there is really there, regardless of whether any person or any other sentient being is looking. It's there 'from its own side,' and it has properties that are its own, regardless of any relation it may or may not have with anyone or anything—properties such as its position and its shape."

"That's right! Now, Julie, the violations of Bell's inequalities mean that we have to give up the paradigm of local realism. So let's give up locality and stop at that. This appeals to me because, as we saw in our last conversation, you only need to assume that what we called 'influences' travel faster than light. We can let the light barrier keep its hold on 'signals.'"

Julie was getting impatient. "What's the matter with you? Why are you so timid?"

Peter had heard the question before from some of his other women friends, but he figured that Julie did not mean it in the same sense they did. Still, he was becoming defensive.

"I'm just trying to minimize the damage," he finally said. "When you have to give up eminently reasonable assumptions, you try to give up the least you can. I mean, I really want to keep both, locality and realism. Giving up locality seems to me the lesser of the two evils."

Julie was glad that not letting her short temper fly was part of her astronaut training. "No matter what happens, keep yourself calm and composed," the old man used to say. She took a deep breath and said in a calm, measured voice, "You know, Peter, I took a course in Western cosmology in my senior year in college, the kind of course that teaches you about Greek cosmology, medieval cosmology, Renaissance cosmology, and so on. There is only one thing I learned from that course, and it is this: Whenever there was a paradigm shift, the old paradigm went with a bang, not with a whimper. When the paradigm of the flat earth went, they didn't try to replace it with an 'almost flat earth,' a flat earth with spherical bumps here and there. If you were in charge back then, you would have said, 'The flat-earth paradigm works so well—let us keep the earth as flat as we possibly can.' Now, as far as local realism is concerned, it is not a question of giving up either locality or realism."

"Really" Why not?"

"I've just been talking to our friend Shimon Malin, and he told me that detailed investigations of Bell's inequalities and their violations lead to the conclusion that locality has to go, regardless of what happens to realism. I believe, however, that once you let a paradigm go, you'd better really let it go."

Peter became pensive. They watched the sunset. Peter enjoyed looking at the big, red wheel slowly sinking into the ocean; then his mind became active again. "Strange," he thought, "I *see* the sun moving, sinking into the water, and yet I *know* that it is the earth that is moving and not the sun. Had I lived in Aristotle's Athens, and had I listened to his lectures, I would have both *seen and known* that the sun moves." He finally turned to Julie and said, "so you really believe that the paradigm shift which started with quantum mechanics, Bell's inequalities, and all that is a big one."

"Yes, I do." Julie too was pensive.

"Well," Peter responded, "I just find it hard to believe. But there are only two possibilities. If you're dissatisfied with my attempts to get rid of locality and stop at that, there's just one way to go—take the plunge!"

"We have been blind for too long," Julie said, "and by 'we' I mean Western civilization. It is obvious to me that realism is as obsolete as the flat earth."

3. The Fallacy of Misplaced Concreteness

That same evening, after dinner, Peter became rather agitated. Julie noticed his restlessness and invited him for a walk along the beach. Peter did not notice the breeze, the moonlight on the water, or even his companion. After a few minutes he could no longer contain himself.

"Look, Julie," he began, "it is one thing to say, in the abstract, 'I want to give up realism,' and quite another to really give it up. You see the silhouette of the building a little way in front of us? It's the same building we looked at this afternoon."

"Yes, I see. What about it?"

"It is obvious to me that the building is there, in that place, now, and that its being there has nothing to do with me looking, with you looking, or with anything related to consciousness. It is just there. Period. Isn't that obvious to you?"

Peter expected a long-winded reply, full of profound philosophical terms. He was quite unprepared for the answer he heard: "Yes, Peter. It is obvious to me."

He stopped in his tracks, bewildered. "But . . . but . . . if that's the case, you haven't given up realism!"

Peter's confusion aroused in Julie a mixture of amusement and compassion. She found she rather liked him, especially when the facade of a young man who knows what he is about crumbled and she saw in front of her a bewildered little boy. She decided to give him a hand.

"Look," she said, "when one contemplates a change of paradigm, the fact that one is convinced of something or other, the fact that something is obvious, means nothing, *because the old paradigm is the basis of one's mental functioning.* It is on the basis of the old paradigm that one feels 'I am absolutely convinced that . . .' or 'This is out of the question' or 'This is obvious.' Imagine us meeting in a previous incarnation, say, four hundred and fifty years ago. I am telling you about a new theory—Copernicus's heliocentric theory—and you respond, quite incredulous, 'But look at the sunset! Isn't it obvious that the sun is moving?'

"In addition, even if the paradigm one is convinced of proves to be extremely trustworthy in the sense of having an impressive predictive power, that is, even if predictions made on its basis come true, time and again, even then it may be proved wrong and be superseded by another paradigm.

"Take the Newtonian concept of physical reality, the so-called clockwork universe. The discoveries that were made on the basis of Newton's laws are extremely impressive. Moreover, our buildings, highways, and bridges are all built on the basis of Newton's laws. Every time I cross a bridge I trust in Newton's laws, and I have not been disappointed yet! Still, Einstein and the other founding fathers of quantum mechanics proved that the Newtonian concept of nature is all wrong. Space and time are not what he thought they were, gravity is entirely different from the Newtonian concept, strict determinism went overboard, and so on, and so forth.

"So—yes, it is obvious to me that there is a building over there, and that its existence, shape, and position have nothing to do with consciousness. But I know not to trust this feeling of 'It is obvious.'"

As Peter listened to Julie he felt that what she said had substance and could not be easily dismissed, yet he was dissatisfied. He focused all his attention on this feeling of dissatisfaction, until its source surfaced, as if by itself: "Well, Julie, if you don't trust your convictions, you don't trust the obvious, *what do you trust?*"

"I trust my capacity to recognize an abstraction when I see one."

"What?!"

"Look, Peter," Julie responded, "we're getting bogged down in arguments about the existence and non-existence of things because we fail to make an essential distinction, the distinction between abstractions and facts. We always think in terms of abstractions, and that's just fine as long as they are recognized as abstractions and their domain of validity is delineated. If one either takes abstractions for concrete facts or applies them outside their domain of validity, or both, the consequences can be disastrous.

"Let me give you some examples. Here's one: 'The human body is a mechanical device.' Very useful abstraction for fixing broken bones, calculating the pressure on different parts of the body, and so on. This abstraction is innocuous as well as useful, because we all know that the human body is not really just a mechanical device. For example, when the capacities of the brain are investigated, the abstraction 'The human body is a mechanical device' is all but forgotten.

"For investigations of the brain another abstraction kicks in: 'The brain is a computer.' Now the situation is neither quite so clear nor quite so innocuous. There is no question that this abstraction has proved very useful in brain research, as well as in computer research—'artificial intelligence,' that sort of thing. The situation is not so clear, however, because the domain of validity of the abstraction has not been well delineated.

"Now we can turn to abstractions that have had disastrous consequences. Take the abstraction 'The earth does not move.' We all use this abstraction, implicitly, all the time. We don't consider the movement of the earth in the vast majority of our activities. It is irrelevant. When an architect plans a house, he or she does not think of the piece of land on which the house will be built as moving. The reason this incorrect abstraction doesn't cause us any harm is that because we know the domain of validity of the abstraction. For example, in calculations of global weather patterns the movement of the earth around its axis *does* have to be taken into account: The assumption of an immobile earth will give entirely wrong weather predictions.

"Back in Galileo's time, however, people had a different relationship to this abstraction. It was not considered an abstraction but a concrete fact—and church dogma, no less. As a result, Giordano Bruno was burned at the stake, and Galileo had his own well-known dealings with the Inquisition. This sad story is essentially a story of mistaking an abstraction for a concrete fact."

"I see what you're driving at," Peter said. "You mean to tell me that realism is an abstraction, that my statement 'There is a building over there, existing on its own, regardless of acts of consciousness' is an abstraction rather than a concrete fact."

"That's right."

Peter did not even try to hide his frustration. "I don't get it," he finally said. "If my statement 'There is a building over there' is an abstraction, what is a concrete fact? Give me an example!"

"Your concrete fact at this moment is 'I see a building over there.'"

"But that's what I said!"

"Not at all. You said, '*There* is a building over there.' This is an abstraction. Your concrete fact behind the abstraction is that you see a building over there. You can question whether there is or there is not a building over there. You cannot question the fact that you are having the experience of seeing one."

Peter's frustration continued to mount. It was his turn to keep himself calm and steady. "Come on, now! You don't really doubt that there is a building over there!"

Julie did not budge an inch. "Of course I don't doubt it. But didn't we establish just a few minutes ago that what I don't doubt, what I am convinced of, what is obvious to me, is all irrelevant? Do we have to keep going in circles?"

With some mental effort Peter had to admit that, at least logically, Julie had a point. After some thought he pulled out his last and, to his mind, winning card. "All right," he said, "if there is no building over there, what is causing the concrete fact of my seeing it? If you can give an explanation that is even remotely possible, I give in."

Peter was too preoccupied with his own thoughts to notice the mischievous smile that appeared on Julie's face. She stopped in her tracks and said: "This morning, when you were sunbathing, I took a walk along the beach. About fifty yards away from here, further down, in the direction of that rather innocuous-looking building, I saw some people moving around and setting in place a strangely shaped piece of plastic. They looked more like scientists than like construction workers. As it turned out, they were scientists, and the head of the team turned out to be a college classmate of mine. I asked him if he could tell me what was going on. He pulled me to the side and said, 'Julie, please don't tell anyone. It's supposed to be a secret. We are erecting a huge hologram of a building, and when we are finished we are going to conduct a statistical study and find out how many people mistake the hologram for a real building.' So you see, Peter, there really is no building over there. It is a hologram you are looking at!"

Peter's jaw dropped—and then he started laughing so hard that he attracted the attention of a small crowd. "Well done, Julie," he finally said. "I don't remember the last time my leg was pulled so successfully."

"Thank you, Peter. For a second or two you really doubted that there was a building over there. Right?"

"Absolutely."

"Well, you asked for possible ways we could be seeing a building over there when there is no 'real' building. I gave you one scenario. Here's another. We went to a show last night, and the show featured a hypnotist. You don't remember anything about it because you were hypnotized, and a post-hypnotic suggestion was made to the effect that you would forget the whole show. Another post-hypnotic suggestion was that you would feel like taking a walk with me after dinner and would see a building where there is none. I asked the hypnotist to give you that suggestion."

"If you continue at this rate, Julie, it will be hard to convince me that there *is* a building over there."

"Good. Next, and the simplest of all: You may be sound asleep and dreaming. When we are dreaming, whatever appears in our dreams seems to have an independent existence but is totally dependent, of course, on the mind of the dreamer.

"If the dream suggestion doesn't appeal to you, let's take it one step further. Maybe Shakespeare was really on target when he wrote, 'We are such stuff as dreams are made on.' The more you contemplate the possibility that *we are dream figures in some higher being's dream*, the less easy it is to dismiss it. If this were the case, we would act, think, and feel precisely as we do now, and yet there would be no independently existing reality. So your concrete fact 'I see a building over there' *can* occur without the ensuing abstraction 'There is a building over there.'"

"Well done, Julie. My compliments, really."

Julie was not displeased. She appreciated her companion's honesty and sincerity. But she was not finished. "Wait, Peter, I haven't told you the best part yet. I told you all kinds of possibilities that my limited mind could come up with, based on what I have studied and read. But the range of possibilities that I know about is really minuscule compared to the range of possibilities that are unknown. That's why I have no problem at all admitting that there is a real possibility that the nature of the physical world is entirely different from realism, even when no alternative paradigm is available to me as yet. At any rate, realism is an abstraction, and all abstractions have domains of validity. As long as this domain has not been delineated, we are likely to apply realism where it is invalid. The result will be mistakes. Whether the mistakes will be innocuous or catastrophic—that's hard to tell."

"I see," Peter responded. "We started our last conversation by asking for the meaning of Bell's discovery and Aspect's experiment. You are saying that realism is consistent with Newtonian physics, but when we come to quantum mechanics, realism is suspect—it may well be an untenable abstraction. Since, obviously, the insights into nature provided by quantum mechanics supersede those of classical physics, the realistic picture of nature—separate bits of matter, existing 'from their own side' at definite locations and with definite shapes—may not represent our best understanding. From the point of view of our current best understanding, realism may be plainly wrong."

"That's exactly what I mean."

"Well, if we give realism up, what will we replace it with?"

The sound of Julie's merry laugh rebounded among the rocks. "That's enough for one evening," she said. "Let's go have a drink."

"How could I possibly refuse?" Peter responded with a broad grin. "I'll drink to concrete facts and appropriate abstractions." After some hesitation he added, "Julie, when you accused me, a little while ago, of being timid, did you have in mind only my interpretation of quantum mechanics?"

4. The Concrete, the Abstract, and Bell's Inequalities

Peter's question "If we give realism up, what will we replace it with?" will be the subject of later chapters. Here we are trying only to establish the difference between concrete facts and abstractions, as well as the fact that realism is an abstraction. "The fallacy of misplaced concreteness" is the term coined by Whitehead for mistaking an abstraction for a concrete fact. As he pointed out, this fallacy is a mistake that derailed Western philosophy.

Whitehead's point is this: When we accept the realist position, we feel that we cannot deny the "fact" that objects exist "from their own side," independently of consciousness. On the other hand, we cannot deny that we do have experiences; i.e., we cannot deny the existence of mind. So the stage is set for the insoluble matter-versus-mind dichotomy. Which is the more fundamental principle, mind or matter? Is one of them real and the other derivative? How do the two interact? A slew of unanswerable questions, unanswerable because the conceptual framework in which they arose is all wrong, all based on the fallacy of misplaced concreteness.

If the concrete fact is not the independent existence of objects, what is it? As Julie explained in the previous section, it is the *experience* that is concrete. Let us expand on Julie's statements by analyzing a concrete fact. I see a building over there. Where is this fact taking place, and what is the relation of the fact to the presumed location of the building?

Where is the fact taking place? It is my experience. It is taking place right here, where I am. But is the building right here? No, of course not. As Julie explained so well, the building may not even exist. How does the place *where the building seems to be* enter the experience? It enters it as a *place of reference*. My experience, which is here, where I am, *has reference* to the place where I see the building. I see the building *in the mode of having location where it seems to be*. Does it follow that something is happening at that place of reference? Not at all. The question of whether something is happening there or not is a separate issue. All we are trying to do now is be very clear, first, about what is concrete and what is abstract and, second, about the location of the concrete and the place or places it refers to.

Whitehead clarifies this point by means of an example. Suppose you are standing at location A, looking in the mirror, and seeing some green leaves that belong to a plant behind your back. Then:

For you at A there will be green; but not simply green at A where you are. The green at A will be green with the mode of having location at the image of the leaf behind the mirror. Then turn around and look at the leaf. You are now perceiving the green in the same way you did before, except that now the green has the mode of being located at the actual leaf.[1]

How does the fallacy of misplaced concreteness apply to the enigma of Bell's inequalities? In the derivation of Bell's inequalities one assumes the independent existence of two particles having their own properties, which include all three spin components. The assumption of realism, which is an abstraction even when applied to people, buildings, and cars, is certainly of dubious validity when applied to subatomic entities.

Can the realization that realism is an abstraction shed light on Bell's correlations? The distinction we arrived at in Section 7 of Chapter 7 between signals which cannot propagate faster than light, and influences, which can, gives us a clue. Every abstraction has a domain of validity. When we say that a signal propagates from an "emitter" to a "receiver," we are using the language of realism. This language seems appropriate to situations involving measurements in the domains of classical physics and Special Relativity. In analyzing such measurements, realism is an appropriate abstraction, and the principle "Nothing moves faster than light" is valid. When we come to influences, however, the situation is much less clear. It may well be that "something" *seems* to propagate faster than light because what is going on is not described in the proper language; the abstraction of realism no longer applies. If this is so, then the difficulty of understanding the significance of Bell's correlations is due to the application of the abstraction of realism outside its domain of validity. As we will see in the next section, this is precisely the message one can deduce from Bohr's framework of complementarity. And we will come back to Bell's inequalities in Section 6 of Chapter 16.

5. The Perspective of Complementarity

It is natural to ask, "If the description based on the assumption of realism is invalid, what is the alternative, and presumably correct, description?" If this question is posed in relation to the quantum domain, then Bohr's framework of complementarity casts doubt on its validity. Bohr claims that in the quantum domain the realistic description of classical physics is not replaced by another, "correct" description; it is replaced, rather, by two descriptions, having different domains of validity. In particular, Bohr makes a distinction between a description that applies to the system when it is isolated—not being measured, not in interaction with other systems—and a description that refers to the quantum system when it is interacting with a measuring apparatus.

In the first case, when the system is isolated, its description cannot be validated. Therefore any description we attach to it is speculative; it is *pure abstraction*. We can go even further: The very existence of the system when thus isolated is a pure abstraction. It is a necessary abstraction, if we are to come up with a description at all, but it is an abstraction, and not a concrete fact.

It is remarkable that, simultaneously and independently of each other, Bohr and Whitehead arrived at this subtle understanding of the role and limitations of abstractions. Abstractions are both necessary and risky. One needs to keep in mind that they are not concrete facts.

In the second case we are at the other extreme: The system is in the process of being measured. During a measurement one cannot separate the measured system from the measuring apparatus. They become one "whole phenomenon."

Bohr's reasoning leads to the conclusion that the two complementary modes of description do not refer to the same quantum entity. The first describes an abstraction, namely, an isolated system whose properties and very existence are, in principle, unknowable; the second describes an interaction in which the interacting parties are inseparable. At any rate, the perspective in which the particles in the EPR or the Bell set-up, particle A and particle B, are little things having their locations and their properties has disappeared without a trace. It is unfortunate to continue calling them "particles," since particles they are not, but we don't have a better word.

In this chapter we have begun to move away from realism and toward the new paradigm indicated by the seemingly strange features of the quantum theory. Our first step has been the realization of the fallacy of misplaced concreteness: By taking the existence of objects in space and time as a primary datum we mistook mental constructs for independently existing entities; we mistook the abstract for the concrete. (Further powerful arguments against realism are introduced in Chapters 9 and 11.)

As Peter pointed out at the end of his conversation with Julie, this realization, while debunking realism, does not provide us with an alternative. The search for such an alternative world-view is what the remainder of this book is about. Chapter 9 is devoted to the next step in this search—an understanding of the process whereby, unawares, we make this mistake of imbuing our mental constructs with an apparent independent existence.

9. Objectivation

Schrödinger's analysis of the process of "objectivation" complements Whitehead's analysis of "misplaced concreteness." Schrödinger points out that the first step of a scientific investigation is the removal of the "Subject of Cognizance" from the domain of inquiry. This removal makes the subject matter of the inquiry a lifeless object. Because this is done universally and unconsciously, it leads to such disastrous consequences as the scientific view of the world as "colorless, cold, mute" and the mind/matter dichotomy.

~

Dear reader or, better still, dear lady reader, recall the bright, joyful eyes with which your child beams upon you when you bring him a new toy, and then let the physicist tell you that in reality nothing emerges from these eyes; in reality their only objectively detectable function is, continually to be hit by and to receive light quanta. In reality! A strange reality! Something seems to be missing in it.

—Erwin Schrödinger

1. The "Hypothesis of the Real World"

In the latter part of his life Erwin Schrödinger published a number of books on a variety of subjects: the General Theory of Relativity, cosmology, thermodynamics, philosophy, the relationship between science and humanism, and even biology. His book *What Is Life?* was influential among both physicists and biologists and played an important part in the process that led to the discovery of DNA.

One of the lesser known among these late books is of particular interest in relation to the issue of realism. Appearing in 1958 under the title *Mind and Matter*, it contains the text of the Tarner Lectures, which were delivered for Schrödinger at Trinity College in Cambridge, England, in October 1956. (Schrödinger could not go to England to deliver his lectures because of ill health.)

The third lecture is entitled "The Principle of Objectivation." It is devoted to a presentation and discussion of an *unconscious assumption* that underlies our perceptions, as well as Western philosophy and science. What is this "principle of objectivation"?

By this [i.e., by "the principle of objectivation"] I mean what is also frequently called the "hypothesis of the real world" around us. I maintain that it amounts to a certain simplification, which we adopt in order to master the infinitely intricate problem of nature. *Without being aware of it and without being rigorously systematic about it, we exclude the Subject of Cognizance from the domain of nature that we endeavor to understand. We step with our own person back into the part of an onlooker who does not belong to the world, which by this very procedure becomes an objective world* [italics added].[1]

This passage is clear, precise, and concise. The reader cannot doubt that Schrödinger knows what he is writing about. But what does it mean?

I discussed this question with my friends Julie and Peter. The three of us were sitting at a coffee shop, cups of steaming coffee in front of us; Julie ordered a glass of water as well. I shared with them Schrödinger's statement of the principle of objectivation. After pondering the statement for a few minutes, Peter turned to Julie.

"You know, this passage seems to take us a step beyond our last discussion," he said. "What Schrödinger calls the 'hypothesis of the real world' is, more or less, what we called in our last conversation 'the abstraction of realism.' He says that we adopt this hypothesis because it is a 'simplification' that helps us 'master the infinitely intricate problem of nature.' Well, that's pretty much the reason behind any abstraction: When we can't deal with the whole situation, we *abstract* what seem to be the most important parts and forget about the rest. This takes care of the first sentence in this quote. The next two sentences go beyond that, though— and that's where I get confused. What does he mean by, 'we exclude the Subject of Cognizance from the domain of nature that we endeavor to understand'?"

Julie was quiet for a moment. Then she responded: "Well, before we try to answer that, let's backtrack. Does it make sense to you to think of 'objectivation' as a process?"

Now it was Peter's turn to be quiet. After a while he said, "No, it doesn't. I understand what an 'object' is. This cup of coffee, steaming here on the table, is an object. I perceive it as existing right here, in front of me. I understand what a 'subject' is. I, the one who perceives this cup of coffee, am the subject. But this expression 'the process of objectivation' sounds strange to me. You can even make a verb out of it, can't you? You can say, 'I objectivize,' right? Well, what kind of activity could that be?"

"It would be the activity of creating the world of objects around us," Julie responded.

"Creating the world of objects around us!?" Peter sounded incredulous. "What do you mean? I see this cup of coffee, I don't *create* it!"

"Are you sure?"

"Oh, please, Julie, really! Look, sitting down, standing up, lifting my cup, and drinking my coffee are just things that I do. 'Creation' is just not the word for it.

I mean, when I look at this cup of coffee, I'm just seeing it the way it is—I don't *create* it!"

Julie just repeated, an impish smile on her pretty lips, "Are you sure?"

"Of course I am!" Peter felt, however, that there was no point in just getting mad. He stopped for a moment, then resumed, feeling rather pleased with himself: "Look here, Julie. Whenever I do something, or create something, I am aware of either doing it or creating it, right? Here, I am lifting my arm. I do that, and I feel myself doing it," He actually lifted his arm. "Now I look at this cup of coffee. I am aware of looking at this cup, but not of *doing* anything to it, let alone *creating* it."

Julie felt that she had finally heard something that deserved a straight response. "It is not true that while you're doing something, you are always aware of doing it. Just a few minutes ago you told me that as you were driving on the Interstate, you had a lapse of awareness between Exit 3 and Exit 8, the exit you took in order to come here. You remembered the moment when you came back to yourself, while taking Exit 8. 'It was like I had been asleep and just woke up,' you said. This actually happens all the time. Past Exit 3 you were doing lots of things—keeping the car on the road, changing lanes, not to mention looking at the pretty women drivers in the cars you passed—all without knowing that you were doing them. This is just one example of doing without awareness. Believe me, there are lots of others."

"I'll be damned!" Peter found it hard to hide his amazement. "That's just like something out of that passage of Schrödinger that we're discussing: 'Without being aware of it and without being rigorously systematic about it, we exclude the Subject of Cognizance from the domain of nature.' It's actually written here in black and white, 'without being aware of it'! So he's pretty much saying that objectivation, whatever it is, is something I do without even being aware of doing it!"

Julie was a little ahead of him. She said, "Perception is the creation of a mental construct."

"Okay, please run that one by me. I promise I won't get mad this time."

"All right, start by looking at this spoon." Julie lifted up a spoon. Both Peter and I were looking. "It's a whole, unbroken spoon." We felt a little silly, but we nodded. "Now I put it in this glass of water." She did. "What do you see?"

Once Peter took on the role of an obedient student, he was able to maintain it for a while. "I see a broken spoon."

"But this is a whole spoon. Why do you see it broken?"

"I can explain that, that's elementary physics." Peter felt himself on solid ground. "Part of the spoon is in water and part in air. The rays of light that come to our eyes from the part that's in water change their direction when they reach the surface of the water. That's what's called 'refraction.' So when these rays of light that start in the water reach our eyes, they seem to be coming from a different direction, not the direction they *actually* came from. But I am sure that's not the point you're trying to make. What is your conclusion?"

"My conclusion," Julie said, "is that you don't see the spoon. What you see is your *mental construct* of the spoon, a construct created by your mind on the basis of the directions from which the light rays arrive at your eyes. If you actually saw the spoon, you wouldn't have seen a broken spoon but a whole one.

"By the way," Julie added, "neurologists who researched the visual cortex, the part of the brain in which images, such as the image of the spoon, are created, have found that only about half of the information coming into this part of the brain comes from sensory impressions. The other half comes from other parts of the brain! The visual cortex is far from being a passive recipient of sensory input. In fact, it's very active. It uses all kinds of input to construct the image of the spoon that you see in front of you. *This activity is as close as our science can get to describing the process of objectivation.* The secret is—and it's quite a secret—that this process takes place totally outside the field of your awareness. Schrödinger was very precise when he added the phrase 'without being aware of it.' What we *are* aware of is the final product, and the final product only—this stupid spoon." She was on a roll, but she stopped as suddenly as she had started.

"So that's what Julie meant when she said, 'Perception is the creation of a mental construct,'" Peter thought. He found the idea a little hard to accept and was trying to explore its implications. "I don't see this spoon; what I see is an image that I construct. As far as seeing goes, each of us lives in his or her private world of visual images . . ."

"When you look, how is a basketball different from the moon?" Julie was back at center stage.

"Are you all right? You need a glass of water, a martini, or something?"

"No, seriously, when you look, how is a basketball different from the moon?"

"All right, Julie. You know that I give you more credit than I give anybody else." He was tempted to add "more credit than you deserve" but bit his tongue. "I'll play along with you on this one. A basketball and the moon . . . well, there are lots of differences."

"What I mean is this," Julie felt she had to explain. "Suppose you look at a basketball from a distance, so that it seems to be about as big as the moon. Forget about the differences in the little details you see on either. What is the major difference in your perception?"

We were all quiet, as Peter and I were thinking. Finally Peter's face lit up. "A basketball looks like a sphere. The moon looks like a flat plate."

Julie beamed. "Right! Next question: Since both the basketball and the moon are really spheres, why does one look like a sphere and the other like a flat plate?"

Now it was Peter's turn to feel that he was on a roll. "That's because the light rays from the moon are exactly the same as the light rays from a flat plate. There is nothing in the configuration of the light rays to indicate a spherical object."

"True enough," Julie agreed, "but isn't it the same with the basketball? If so, why do they look different?"

We were quiet again. Finally Peter said flatly, "I don't know."

"All right, I'll tell you. The reason for the difference is that whenever we play basketball we use a *basketball*, we never use the moon."

"Of course not," said Peter, exasperated, "but what does this have to do with anything?"

"Actually, a lot. Take a look at it this way: Our only experience with the moon is visual, and, as you've just explained, as far as visual experience goes, there is nothing to indicate a spherical shape rather than a flat one. With a basketball, however, it's different. We not only look at it, we also touch it; and, touching it, we know it's round. This experience of touch is included in the information that the brain uses when it constructs the image of a basketball. This simple example shows that what's going on in the brain is real creation, the creation of mental constructs. These mental constructs are perceived as if they exist as actual objects, rather than as what they are—as mere mental constructs."

"Come to think of it," Peter responded, "I've heard a different explanation. We see the depth in things with the help of the stereo effect of the eyes: The eyes are situated in slightly different locations; therefore their perspectives on the things we see are slightly different. We create a three-dimensional image by combining and integrating these two perspectives. This is easy to do with a basketball, but the moon is so far away that there is no perceptible difference between the perspectives of the two eyes."

"This may be true, but it's only a part of the story," Julie said, "When we look at round, earthly objects that we know are round, we see them as spherical, even when they are too far away for the stereo effect to be effective. A big balloon seen from a distance is one example. And, even if we accept your explanation, the integration of the two perspectives of the two eyes into a single, coherent image is still no mean task! By the way, it takes us a long time to learn how to create these images, these mental constructs. For two or three years we were very busy learning how to do it."

"Really? What do you mean?"

"I mean the first two or three years of our lives. The Swiss psychologist Jean Piaget studied this process of learning in great detail. For example, until a baby is about eighteen months old or so, his or her perception is confined to two dimensions. The baby doesn't see depth. It hasn't learned that yet."

"Yes, I have verified that myself," I interjected. "When my oldest son was a baby, I was very much into Piaget, so I watched him closely with Piaget's findings in mind. It so happened that this baby stood up and walked by himself very early, at eight or nine months. At the time, we lived in an apartment that had rooms on two levels. When I was upstairs with him, it was scary: He would walk toward the stairs ready to step on them, as if they were painted on the floor. He clearly did not have a perception that included the possibility of falling down these stairs."

"So," Julie concluded, "perception is the creation of a mental construct. The capacity to be creative in this way is hard won. Babies work hard to acquire it. Once we have this capacity, we use it all the time, and we're totally unaware that we do it. What we call 'external reality' is something we create, we objectivize, and yet we believe it is simply there. The word 'objectivize' is appropriate—we create the *objects* we see all around us. Now we're ready to address Peter's question."

Peter was caught off guard. "What question?" he asked.

"Why, the very question that started our whole conversation: What does Schrödinger mean by 'we exclude the Subject of Cognizance from the domain of nature that we endeavor to understand'?"

2. The Exclusion of the "Subject of Cognizance"

It was getting late. We got up and started strolling along the sidewalk, objectivizing shop windows, passing people, and cars, as well as each other. After a few minutes Julie spoke: "Your question, Peter—are you sure you don't know the answer already? It is so obvious, so basic, that I find it hard to explain. Let me see. . . . When we look around, we see this external world nicely ordered in space. In this vast space, filled, as it were, with objects and void, where are you?"

"Me? I'm right here!"

"No, your body is here. Where are you?"

Once again Peter started to feel as if the ground were slipping from under his feet. "Where am I? Inside my body, I suppose."

"Really? Where? Certainly not in your toenails! Not even in your arms or legs. They can be amputated and you will still be here. Not even in your heart. You can have a heart transplant and you will still be here."

Peter found these images unpleasant to contemplate. He tried to defend himself from Julie's assault and threw out an answer that he immediately regretted. "I am in my brain!" he declared, now feeling a little foolish.

Julie was relentless. "Where in your brain?"

"How should I know? I can't even name the parts of the brain. The neurologists should know that, especially the neurosurgeons."

"So, when you undergo brain surgery you expect the surgeon to find you somewhere in the deep recesses of the brain."

"I see what you mean. No matter where you look, legs, heart, brain, all you see is bones, tissues, neurons—you never see *me* or *you*."

"That's right. How *could* you see me or you? Whenever you see, you see objects. As far as the question of what you are is concerned, one thing is certain: You are not an object. You are the *subject*. This space that is all around us is a space of objects, and you, the subject, will never be found in it, because you are not an object. This exclusion of you and me is an integral part of the process of objectivation. When we objectivize, we do two things, both at once (it's really one process, but with two aspects): We construct this world of objects, and we exclude the 'Subject of Cognizance' from it. Schrödinger put it very well: 'We exclude the Subject of Cognizance from the domain of nature that we endeavor to understand. We step with our own person back into the part of an onlooker who does not belong to the world, which by this very procedure becomes an objective world.'"

"I almost see what you mean, but I need just a little more help."

Julie became thoughtful. "There's not much more I can do by way of explanations," she finally said. "This idea of objectivation is not only profound, it is also very simple. If we keep trying to explain it, we may end up being caught up in words. But I can do something else for you. I can give you an analogy."

"Let's hear it!"

"You and I are characters in this book that Shimon Malin has written. Shimon objectivized us as these characters. Right?"

"Yes, of course. We even have the pleasure of being with him; he's walking here next to you. He doesn't say much, but he's here. 'Meet your author,' as they say in those literary ads."

"No, you got it all wrong. This Shimon Malin who is walking next to me is not the author. He is one of the characters. Like us."

"I don't get it."

"Look. Why is Shimon walking here, with us? Because the author, back at the beginning of the previous section, decided to write, 'The three of us were sitting at a coffee shop.' He could have left himself out. He could have written, 'Julie and Peter met in a coffee shop.'"

"He must have had his reasons."

"I don't doubt it, but that's not the point. The point is that this Shimon who's walking next to me is as dependent on the author as we are. And Shimon Malin, the author, is nowhere to be found *in the book*. He is like the 'Subject of Cognizance' that cannot be found in the objectivized world."

"I follow that. It's like when you dream. You, the dreamer, objectivize the whole dream, and you also take part in the dream as one of the dream figures—one person among many. So in the dream situation there are two of you: the dreamer and the dream figure. The two have the same name. Both are Julie. One Julie is the dreamer, lying in bed, dreaming; the other is the Julie in the dream, talking to people, going to a party, drowning, fighting, or whatever. The first Julie is not in the dream. The second Julie is not dreaming."

"Right. Now, in the dream situation, I, as the dream figure, am liable to make enormous efforts to get something done or to avoid some imminent danger. If I knew I was the dreamer, I could change the dream easily, any way I choose."

"So that's my situation," I said. After my long silence Julie and Peter were happy to hear me speak. "I can work hard to achieve something as a character in the book, or else *he* can get it done easily as the author."

Julie smiled and nodded, but Peter wasn't sure what I meant.

"Look, Peter," I said, "suppose I am attracted to Julie, and I wish to be alone with her at this point in the book. I can do one of two things: I can try to get you out of the way, saying, perhaps, that it is getting late and it is time *for me* to take Julie home. You will say, "Sure, let's *both* accompany her home," at which point I'll feel stupid and frustrated, or—"

"Or what?" Peter was annoyed with me. He didn't like my example.

"Or, *he* can sit at his desk and write: 'As they were walking, Peter remembered, all of a sudden, that he had an important engagement. Hurriedly he took his leave. Julie and Shimon, having waved good-bye to him, looked into each other's eyes. They knew then, at that instant, that they loved each other . . .' See? It's all done! Not only is Shimon left alone with Julie, he is even assured that she is as attracted to him as he is to her!"

It was Peter's turn to smile. "Yes, but in this case this mutual attraction is not very meaningful for him, is it? Shimon, the author, has created a passionate attraction between two of his characters. Big deal. That's very different than being in love with Julie, who is right here!"

"Yes," I admitted gloomily, "nothing's perfect."

3. Objectivation: The Disastrous Consequences

Objectivation in itself is not a problem. The problem is that we are unaware of the fact that we objectivize. In particular, we are unaware that "we exclude the Subject of Cognizance from the domain of nature that we endeavor to understand." Being unaware that we have excluded ourselves, we take what's left after this exclusion, namely, the world of objects in space and time, as all there is.

This construction of objects and the exclusion of the self, together with the ignorance of the fact that we do that, work well in the context of satisfying our basic animal needs. However, mistaking mental constructs for independently existing objects in the context of deeper needs, such as the need to know the universe and our place in it, has disastrous consequences.

One disastrous consequence is this: If "the real world" is just atoms in space, then values and meanings are not real. Even some attributes of objects, such as color or warmth, are not real; they are merely the way one's sensory apparatus interprets the impact of atoms on it. "The real world" is not only meaningless but also, in Schrödinger's words,

"colorless, cold, mute." Color and sound, hot and cold are our immediate sensations; small wonder that they are lacking in a world-model from which we have removed our own mental person.[2]

Another disastrous consequence is this strange impasse of Western philosophy, "the mind-body problem." If the objective, external world is really all there is, then each one of us performs amazing feats of magic all day long: I decide to raise my hand, and I raise my hand. How did my decision—a mental event—bring about the physical event of a physical hand moving up? To solve this riddle philosophers (and others) keep looking for the mind-body interface. But mind is not even a part of the objectivized reality, which is, presumably, all there is! On the one hand, the presence of mind cannot be denied—we know that we do experience—and on the other hand, it is nowhere to be found in what we believe to be "the real universe."

Schrödinger summarizes our predicament as follows:

So we are faced with the following remarkable situation. While the stuff from which our world picture is built is yielded exclusively from the sense organs as organs of the mind, so that every man's world picture is and always remains a construct of his mind and cannot be proved to have any other existence, yet the conscious mind itself remains a stranger within this construct, it has no living space in it, you can spot it nowhere in space. We do not usually realize this fact, because we have entirely taken to thinking of the personality of a human being, or for that matter also that of an animal, as located in the interior of its body. To learn that it cannot really be found there is so amazing that it meets with doubt and hesitation, we are very loath to admit it.[3]

4. Beyond the Subject/Object Mode

Whitehead's realization of "the fallacy of misplaced concreteness" and Schrödinger's analysis of "the principle of objectivation" both lead to the conclusion that so-called objects are mere mental constructs. It follows that "the real world," in itself, is not a collection of objects. This is a strong statement about what "the real world" is not, but what insight do we have as to what it is?

At first glance it seems that the quantum theory is not the place to look for the answer to this question. The quantum theory is an integral part of Western science, and, as such, it is subject to the principle of objectivation. Experiences per se have no place in it. Surprisingly, however, it turns out that an understanding of quantum measurements yields crucial indications about the new paradigm we are looking for.

The quantum theory of measurement and its bearing on the emergent worldview is the subject of the next two chapters. We will end the present chapter by quoting the last paragraph of Schrödinger's essay "The Principle of Objectivation." The paragraph may seem obscure, but it suggests the direction we should take in order to find the solution to our inquiry. We will come back to it in Chapter 10, Section 5 and Chapter 19, Section 4.

The world is given to me only once, not one existing and one perceived. Subject and object are only one. The barrier between them cannot be said to have broken down as a result of recent experience in the physical sciences, for this barrier does not exist.[4]

10. In and Out of Space and Time

In this chapter we zero in on the mystery at the heart of quantum mechanics, "the collapse of quantum states." The collapse is the process whereby, in an act of measurement, the potential becomes actual: An "elementary quantum event" is created out of a background of potentialities. The process is atemporal, occurring outside space and time, and leads to the appearance of an actual event in spacetime. The idea of an atemporal process may seem strange, but it is neither new nor obsolete. It is an integral part of Plato's world-view, as well as Whitehead's. The understanding of atemporal processes calls for both discursive reasoning and contemplation.

~

The things which are temporal arise by their participation in the things which are eternal.

—Alfred North Whitehead

1. Elementary Quantum Events

As we saw in Chapter 4, a quantum measurement brings about a transition from the potential to the actual. An isolated electron is a "field of potentialities." When it is subjected to a measurement, however, it emerges into actual existence and becomes an "elementary quantum event."

Consider, for example, a measurement of position. If we wish to measure the position of electrons, we can use a TV screen. The impingement of an electron on such a screen results in a flash of light. This constitutes a measurement of the position of the electron because the location of the flash tells us where the electron was at the instant of impingement.

Can we predict where the electron will hit the screen? No, we can't. All we have is a probability distribution—a statement of how probable it is for the electron to arrive at different locations. We know that some locations are favored, but we don't know where it will actually hit. How, then, is the actual place of impingement being chosen? And how does the transition from the potential to the actual take place?

These are difficult questions. Physicists have been arguing about them for seventy years, and no consensus has been reached. In this chapter and the next I will share with you my own conclusions, the result of four decades of study, thought, and contemplation.

Let us begin by considering the simple statement "An electron is moving toward a TV screen." This sentence is fatally flawed. *If the electron does not exist at all as an actual object when it is isolated, how can it be moving toward a TV screen?* The non-existence of electrons, when isolated, was discussed in detail in Chapter 4, and yet, in all likelihood, you read the statement "An electron is moving toward a TV screen" without raising an eyebrow. This is an illustration of the power of paradigms. The paradigm that states that matter is made of particles whose actual existence is continuous is holding us captive: It unconsciously reestablishes itself as soon as we have finished reading that electrons have no actual existence except when measured!

Well, then, if this innocuous-looking statement, "An electron is moving toward a TV screen," is wrong, what is the right statement? Let's give it a go: "Before the event of impingement, there is a field of potentialities for finding the electron at different locations in space should a measurement of its position be carried out. As time goes by, these potentialities change in such a way that it is getting more and more likely that, if measured, the electron will be found closer and closer to the screen. Eventually, when there is a probability distribution for finding the electron *at the screen*, it is indeed found there, as an 'elementary quantum event,' giving rise to a visible dot, because the TV screen itself serves an a measuring apparatus for position."

This sentence is correct, but how cumbersome! It is impossible to say it simply, however, because *our language is steeped in the current paradigm.* This is a difficulty we face again and again. It is not only a difficulty of expression, it is a difficulty of thought as well. Language and thought are bound together, and both pull us toward the current paradigm by forces that are as strong as they are unconscious. Indeed, the attempt to break away from this persistent pull and see at least the beginnings of a new world-view is the aim of this book.

In speaking about an electron that is, or isn't, moving toward a TV screen, the correct ways of expression are so unwieldy as to be impractical. If we wish to move on, we are condemned to state things incorrectly. Let us agree that *the clear and incorrect statements we can't help making will be considered indications for the correct, cumbersome ones.*

To recapitulate: An electron is moving toward a TV screen. Before impinging on the screen, the electron does not exist in space; it is merely a field of potentialities. At the time of impingement, however, it does actually exist; it is an elementary quantum event, an event in space and time. Through the interaction with the screen, which is a measuring apparatus for position, a transition from potential to actual existence has taken place. This transition is called "a collapse of the quantum state (or wave function) of the electron."

The word "collapse" refers to the fact that before the transition, the probability distribution is spread out; the whole screen is available to the electron. Once the electron becomes actual, however, it is found *somewhere* on the screen, and there is no probability of finding it anywhere else. The probability distribution has "collapsed" to a single location.

Having defined our terms, we are ready to probe into the structure and nature of this process of collapse.

2. The Structure of the Collapse

The process of collapse involves three stages. To explain them in context, consider, once again, our electron, a field of potentialities, as it is moving toward a TV screen. The field of potentialities consists of different kinds of potentialities. One kind, the kind we have been discussing so far, is the different possibilities of hitting the screen at various locations, i. e., the different possible results of a *measurement of position*. If, however, instead of measuring position we measure momentum, we evoke another kind: the probabilities of the various values of the momentum that could result from a *measurement of momentum*. We may decide to measure neither position nor momentum; we may decide to subject the electron to a measurement of the x- or y- or z-component of its *angular* momentum. There are also other quantities that can be measured; the field of potentialities contains the probability distributions for all of them.

In the case under discussion, however (in which the electron is moving toward a TV screen), it *is* the position that is measured. In the first stage of the collapse the field of potentialities aligns itself with this type of measurement. The wave function is represented in terms of all possible positions (all possible locations on the TV screen) rather than, say, all possible values of the momentum.

Stage 1: a choice of a particular set of potentialities, which corresponds to the measurement apparatus.

An analogy may help clarify the meaning of stage 1. A few minutes ago I took a break from writing. I was just leaning back in my chair, relaxed; my mind was not focused on anything in particular. All of a sudden I heard my wife's voice. "We have talked about all the places we could go to on our vacation," she said, "We've discussed them long enough, and it is time to make a choice. I don't care where we go, but I do want you to make a decision." All of a sudden my attention became focused; I saw all the places that we had been discussing in front of my mind's eye. My mental gaze roamed through Mexico City, Paris, Kyoto, Jerusalem, as well as more unlikely destinations, such as Antarctica.

Before my wife spoke, my attention was not focused on any one of the different *kinds of possibilities* that existed in my future (where to go for vacation, what car to buy, which books to read, whom to meet, etc., etc.). Having heard my wife speak, however, I became focused on the different possible trips, and on nothing else. This focusing is analogous to the choice of a particular set of potentialities as a result of the presence of a particular measurement set-up. In the example of the electron and the TV screen, the chosen set is the collection of potentialities for different locations on the screen. This choice is induced by the presence of the TV screen. Had the TV screen been replaced by, say, a momentum measurement apparatus, another choice of kind of possibility would have been made.

My attention is focused on vacation places, but I have made no decision yet. Analogously, the electron is facing different possibilities as to where it will hit the screen; other conceivable choices, e.g., how fast to move, are no longer relevant. But the choice of the location that will be hit has not yet been made. This choice is made in stage 2.

Stage 2: a choice of one specific potentiality out of the set of potentialities singled out in stage 1.

At this stage a particular spot, e.g., the top right-hand corner of the TV screen, is chosen, and this is where the electron will impinge. In the analogy, this stage is where I make my choice: We will go to Jerusalem.

So far, everything is still happening in the realm of potentialities. Nothing is happening in the actual world. Choices have been made, but no event has taken place. In the analogy, I have thought about vacation places, I have made a decision, but we haven't gone anywhere yet. The transition from the chosen potentiality to its actualization is the third stage.

Stage 3: an actualization of the choice made in stage 2, i.e., the appearance in space-time of the phenomenon that was singled out.

This is the last stage of the collapse. Now an actual elementary quantum event is taking place: The electron is impinging on the screen at the chosen location in the right-hand corner. This impingement leads to the momentary appearance of a bright dot at that location. The transition from potentialities to an actuality is now complete. In the analogy, stage 3 is the one in which we actually go to Jerusalem.

Naturally, a delineation of the steps in a process is not the same as understanding the process—far from it. The delineation leaves the questions of how and why any of these steps takes place wide open. A correct delineation is essential for providing a framework for asking the right questions, however. It is as true in physics as anywhere else that asking the right question brings one more than halfway to the answer.

We are now ready to start our inquiry into the nature of the collapse. Let us begin with the following question: How long do the three stages in the process of collapse take? The answer to this question may surprise you: Unlike the process of choosing a place for vacation and going there, *the whole process of collapse, including the three stages, does not take any time.* It is an atemporal process, occurring outside of spacetime, whose final outcome is the appearance of an event—an elementary quantum event—in spacetime.

Needless to say, this answer calls for an explanation. What do we mean by this strange concept of an atemporal process?

3. Eternal Order and Temporal Order, or Heisenberg in the Gutter

In his autobiographical book *Physics and Beyond,* Heisenberg recalls the situation in Munich after the end of the First World War:

> In the spring of 1919 Munich was in a state of utter confusion. On the streets people were shooting at one another, and no one could tell precisely who the contestants were. . . . Pillage and robbery . . . caused the term "Soviet Republic" to become a synonym of lawlessness, and when, at long last, a new Bavarian government was formed outside Munich, and sent its troops into the city, we were all of us hoping for a speedy return to more orderly conditions.[1]

Heisenberg joined a volunteer force that was formed to act as guides to the advancing troops. This was the beginning of his military service. Eventually things quieted down, and the military service became "increasingly monotonous."

> Quite often it happened that, after spending the whole night on guard in the telephone exchange, I was free for a day, and in order to catch up with my neglected school work I would retire to the roof of the Training College with a Greek school edition of Plato's Dialogues. There, lying in the wide gutter, and warmed by the rays of the early morning sun, I could pursue my studies in peace.[2]

It was there, in the gutter on the roof of the Training College, that Heisenberg discovered the *Timaeus,* Plato's account of the creation and structure of the universe. Heisenberg was both fascinated and disturbed by it. The detailed account of the geometrical shapes of Plato's "atoms"—the tetrahedrons, cubes, octahedrons, and icosahedrons that represent atoms of fire, earth, air, and water, respectively—did not make much sense to him.

> The whole thing seemed to be a wild speculation, pardonable, perhaps, on the grounds that the Greeks lacked the necessary empirical knowledge. Nevertheless, it saddened me to find a philosopher of Plato's critical acumen succumbing to such fancies. I looked for some principle that would help me to find some justification for Plato's speculation, but, try though I might, I could discover none.[3]

As Heisenberg continued his studies, he came across new questions, inspired, most likely, by the unsettled conditions at the time. The theme of order kept coming up. He had grown up in a world that had the appearance of order. This order was shattered, however, by World War I and, in its aftermath, by clashes between movements aiming at the restoration of the old order and movements proclaiming a new order. Against the background of these conditions, the eternal order, described by Plato, kept attracting him:

I kept wondering why a great philosopher like Plato should have thought he could recognize order in natural phenomena when we ourselves could not. What precisely was the meaning of the term? Are order and our understanding of it purely time-bound?[4]

This last question goes to the heart of the *Timaeus*: What is the relationship between the eternal order and the temporal order, i.e., the relationship between the order in Plato's "intelligible world," the world of Forms, and the order in the visible, corporeal, time-bound universe?

Heisenberg's preoccupation with the idea of order led him to a profound experience. A few months after he read the *Timaeus* he participated in a large gathering in the courtyard of Prunn Castle, which "stands sheer on a rock at the edge of the valley."[5] As he was listening to different speakers proclaiming different types of order, he could not help noticing that the different suggestions clashed with each other in a clearly disorderly fashion.

This, I felt, was only possible because all these types of order were partial, mere fragments that had split off from a central order; they might not have lost their creative force, but they were no longer directed towards a unifying center. Its absence was brought home to me with increasingly painful intensity the longer I listened. I was suffering almost physically, but I was quite unable to find a way towards the center through the thicket of conflicting opinions. Thus the hours ticked by, while more speeches were delivered and more disputes were born. The shadows in the courtyard grew longer, and finally the hot day gave way to slate-grey dusk and a moonlit night. The talk was still going on when, quite suddenly, a young violinist appeared on a balcony above the courtyard. There was a hush as, high above us, he struck up the first great D minor chord of Bach's Chaconne. All at once, and with utter certainty, I had found my link with the center. . . . The clear phrases of the Chaconne touched me like a cool wind, breaking through the mist and revealing the towering structures beyond. There has always been a path to the central order in the language of music, in philosophy and in religion, today no less than in Plato's day and in Bach's. That I now knew from my own experience.[6]

4. Processes Outside of Space and Time

For many centuries the *Timaeus* was highly influential, not only as a philosophical text but as a text in physics and astronomy as well. Even young Heisenberg, reading it in 1919, tried to decipher it in scientific terms. Regardless of what one makes of it as a physics or astronomy text, there is no question that as a philosophical statement it is one of the peak achievements of our civilization.

The whole of Plato's body of work, including the *Timaeus*, is based on the distinction between "that which always is and has no becoming and . . . [that which] is always becoming and never is,"[7] that is, the distinction between "the world of immutable being" (also known as "the world of Forms" or "the noumenal world") and "the world of generation." The universe consists of both; and Plato's account of its creation depicts first the creation of "the world of immutable being" and then the creation of "the world of generation."

"The world of immutable being" has nothing to do with time. It just is. It is not only unchanging but also, by its very nature, unchangeable. The concepts of time and change do not apply to it. Equality, Sameness, Difference, Beauty, Justice, and all the other Forms are immutable—and yet Plato speaks of the process by which they were created! Clearly, he does not mean creation as a process in time. Indeed, we read about the creation of Forms, and of much else besides, before we get to *the creation of time.*

According to *Timaeus*, the creation of the universe begins with the creation of Soul—the soul of the universe. Soul is eternal; it belongs to "the world of immutable being," hence its creation is unrelated to time. This creation is a complex process, involving a number of stages. First:

From the being which is indivisible and unchangeable, and from that kind of being which is distributed among bodies, he [the creator] compounded a third and intermediate kind of being. He did likewise with the same and the different, blending together the indivisible kind of each with that which is portioned out in bodies. Then, taking the three new elements, he mingled them all into one form, compressing by force the reluctant and unsociable nature of the different into the same.[8]

Next comes a series of divisions, involving mathematical ratios that have puzzled and intrigued researchers for the past twenty-five centuries. After that come the creation of circles and the introduction of circular motions, which are subsequently divided up among the planets.

It is not necessary to follow the details in order to realize that the process described by Plato is *a creation of principles, not of corporeal things*; hence time has nothing to do with it. When one looks at it metaphysically, as a fashioning and an interrelating of principles, one is dazzled by the depth of thought and insight it contains.

Only when the world soul is completely formed is the creation of a corporeal world possible:

When the father and creator saw the creature which he had made [i.e., the world soul] moving and living, the created image of the eternal gods, he rejoiced, and in his joy determined to make a copy more like the original, and as this was an eternal living being, he sought to make the universe eternal, as far as might be. Now the nature of the ideal being was everlasting, but to bestow this attribute in its fullness upon a creature was impossible. Wherefore he resolved to make a moving image of eternity, and when he set in order the heaven, he made this image eternal but moving according to number, while eternity itself rests in unity, and this image we call time.[9]

Processes that are outside time are central to Plato's thinking. As far as the account of creation goes, they are the essence of the story; a confinement of the account to creation within time would be unthinkable. This notion of atemporal processes is necessary not only for the big story of the creation of the universe; atemporal processes are, according to Plato, all around us. They are the essence of any process of "becoming." A beautiful object becomes beautiful through its "participation" in the Form of Beauty, and this "participation" is not a process in time. It is, rather, an atemporal process that leads to the appearance of a beautiful object in space and time.

This understanding of the term "becoming" is also important when one thinks about the collapse of quantum states. The collapse is a transition from potentiality to actuality. *Space and time refer to the ordering of things and events in the actual world.* A potentiality, like a noumenon, is neither a thing nor an event; hence it does not exist in spacetime. Therefore the process whereby an electron appears in spacetime as an elementary quantum event is not temporal. Just like Plato's "becoming," it is an atemporal process whose final outcome is the appearance of an event in spacetime.

At this point you may wonder: Why should we take Plato so seriously? Hasn't philosophy progressed? And isn't Plato's notion of atemporal processes at all out of date? As to whether philosophy has progressed, one would not be wrong in answering, "Not much." As Alfred North Whitehead writes, "The safest general characterization of the European philosophical tradition is that it consists of a series of footnotes to Plato."[10] And as to the other question, "Isn't Plato's notion of atemporal processes out of date?" the answer is "No, it isn't." Atemporal processes are one of the cornerstones of a very great achievement of twentieth-century philosophy—Whitehead's own "process philosophy." As we will see, Whitehead's atemporal processes are closely related to the collapse of quantum states.

Let us get back now to young Heisenberg, lying in the gutter on the roof of the Training College, Plato's Dialogues in hand. It is possible that the strong impression that the *Timaeus* made on him freed him from his attachment to a "realistic" view of the universe and enabled him to consider reality an interplay between actual events, coordinated in spacetime, and fields of potentialities, which *refer* to space and time while being outside the spacetime matrix. Heisenberg himself gives us a hint in this direction. He was frustrated by his inability to make sense of the detailed account of atoms in the *Timaeus*. "Even so," he writes,

I was enthralled by the idea that the smallest particles of matter must reduce to some mathematical form. After all, any attempt to unravel the dense skein of natural phenomena is dependent upon the discovery of mathematical forms.[11]

The phenomenal world that we know through the senses is not all there is. Plato wrote extensively about the noumenal world, *which is the being of the phenomenal*. St. Augustine wrote, simply, "There is another reality besides this."[12] Heisenberg was well aware of this other reality, not only through his conviction that there are mathematical forms beyond "the dense skein of natural phenomena" but also through his experience of "a central order" at Prunn Castle. He was thus free to explore, mathematically and conceptually, a notion of reality that is not confined to material objects, actual events, and temporal processes.

5. Discursive Reasoning and Contemplation

We left Julie and Peter walking, late at night, and talking about objectivation. The next morning I found them on the beach, mellow and relaxed. I gave them the first four sections of the present chapter to read. Julie was delighted. Peter's response, however, was cautious.

"I have mixed feelings about all this," he told Julie. "Would you deign to be the embodiment of the Form of Wisdom and clear my doubts?"

"I don't know about being the *embodiment* of the Form of Wisdom," Julie responded, smiling. "I'm not quite that ambitious. But I will try to *participate* in the Form of Wisdom, or, rather, let the Form of Wisdom participate in me. Even that's not easy. Sometimes I feel as though the Form of Stupidity were hovering over my head."

"The Form of Stupidity? Is there such a Form?"

"I don't think so. Forms are supposed to be perfect, and stupidity . . . on second thought, one often sees Perfect Stupidity here on earth, so maybe . . . anyhow, what are your doubts?"

"I was thinking about potentialities. The idea that they don't exist in space and time just rubs me the wrong way," Peter said.

"It rubs me the wrong way too," Julie responded, "but I kind of see why."

"Care to explain?"

"Well, if you insist. You see, whenever I think about anything, any concept, I see it as an object (a mental object, of course) in front of my mind's eye. Here am I, the subject, thinking, over against *it*, the object of my thought. Once I relate to it as a subject to an object, I expect to find a place for it in space; and if it doesn't belong to the external space, the space that is, presumably, common to all of us, I expect to find it in my inner space, the space that is in front of my mind's eye."

Peter thought for a moment. "Okay, that's fair enough," he conceded, "but this can't be the whole story. A potential is a potential for something happening somewhere at some time. It is not unrelated to space and time."

"It is *related* to a location in spacetime, all right," Julie was quick to respond, "but it does not *exist in* spacetime."

"This is too abstract for me. Give me an example."

"Remember the phases of the collapse of quantum states and the going-on-vacation analogy? Once Shimon decided to go to Jerusalem, there is a potentiality of him being there. Where is this potentiality? It is related to Jerusalem, but, most assuredly, it does not exist in Jerusalem. Nothing changed *in Jerusalem* as a result of the arising of this potentiality."

"Yes, of course. Unless and until Shimon arrives in Jerusalem, that is, unless and until there is a transition from potentiality to actuality, nothing is happening in Jerusalem. But something is happening in Shimon. I submit that the potentiality is located in Shimon's brain."

An exasperated look appeared on Julie's face. "If you keep going back on what we established already, we'll never get anywhere."

"What do you mean?"

"Haven't we established that all you find in a brain are cells, neurons, tissues, electrical impulses? You can't even find Shimon's *intention* to go to Jerusalem, let alone the potentiality of his being there! But, luckily, there is no need to get into the mind/brain issue right now. Just think about the potentiality of an electron to hit a particular location on a TV screen. This potentiality, where in space is it located?"

"Why, in the electron! Where else?"

"*But the electron itself does not exist as a thing in spacetime. It is merely a field of potentialities!*"

Peter felt as if he had emerged from a cloud of confusion. He was genuinely grateful. "Thank you, Julie," he said. But he was quick to add, "My next question, however, is not so easy."

Fortunately Julie enjoys challenges. "Terrific. Out with it!"

"Suppose I agree—as I do—that atemporal processes, such as the collapse of quantum states, are an essential ingredient of reality. Still, how am I supposed to *think* about them? Whenever I think about a process, I can't help thinking about it as taking place in time!"

Julie was thoughtful for a long time. Peter too was quiet, and as I watched them sitting there, I felt as if the sound of the waves did not really disturb the underlying stillness. It surged and subsided, emerging out of the stillness and merging back with it. The dichotomy of the temporal and the atemporal did not seem all that sharp.

Finally Julie lifted her eyes and gave Peter a penetrating look. "I'll respond to your question in two ways," she said. "The first is incomplete but easy to understand. The second, the real McCoy, is harder to follow. So let's start with the easy one. Remember that time during our astronaut basic training when the old man started speaking about spacetime as 'a curved four-dimensional continuum?' How did you manage to think about this strange concept?"

"Now that you mention it, I realize that I had the same trouble with the notion of 'a curved four-dimensional continuum' back then that I am having with 'atemporal processes' right now."

"Right. And, back then, what did you do?"

"Well, I started thinking about three-dimensional, rather than four-dimensional, curved spacetime. When this didn't seem to help, I dropped the time dimension altogether and started thinking about a two-dimensional curved continuum. This was easy—the surface of a sphere is such a continuum. So I gleaned some of the properties of a curved four-dimensional continuum from thinking about spheres. Then I started thinking about moving spheres, in order to put the time back into it. In short, while I couldn't get the concept of a curved four-dimensional continuum to appear as a clear 'something' in front of my mind's eye, I played with it and explored different aspects of it until I became comfortable with it. The concept of a four-dimensional curved spacetime is no longer a problem for me."

"Fair enough," Julie responded. "I went through a similar process myself. I'd just like to add one element to your description: While playing around with expanding spheres, two-dimensional surfaces cruising through space, and all the rest of it, I never lost sight of the fact that these were partial aspects of a curved four-dimensional continuum. The continuum itself kept lurking in the back of my mind as that which was being hinted at, but never reached, by all these concepts and pictures I was playing with. Now, when we come to atemporal processes, we can proceed in essentially the same way. Think about processes in time, if you like; but remember that the sequence you create in your mind is not really a time sequence. Try to think of it, for example, as just a sequence, or try to compress all the stages into a single instant, without letting them merge. I know it's hard, and I know that none of these ways of thinking about it is really right. Nevertheless, keep dancing around the concept, and you will find it more and more accessible."

"All right," Peter responded, "I get it. Perseverance furthers. No pain, no gain. The long, winding road. Now—what is the real McCoy?"

"No, no," Julie retorted impatiently, "don't make fun of the long, winding road. It is indispensable, both as an independent approach and as a preliminary step— preliminary to the real McCoy. The real McCoy is the kind of thought with which we can really penetrate to the heart of things. Plato mentions it in the *Timaeus*." She pulled a small book from her bag and read aloud:

First, then, in my judgement, we must make a distinction and ask, What is that which always is and has no becoming, and what is that which is always becoming and never is? That which is apprehended by intelligence and reason is always in the same state, but that which is conceived by opinion with the help of sensation and without reason is always in the process of becoming and perishing and never really is.[13]

"Plato is making a distinction here," Julie commented, "as he does in many other places, between two kinds of thinking: 'intelligence and reason' vs. 'opinion with the help of sensation.' I call them 'contemplation' and 'discursive reasoning.' Discursive reasoning can clarify relatively superficial aspects of issues and concepts: logical relationships among arguments, formal connections among concepts and issues, analogies, and so on. To get to the heart of a deep issue, question, concept, or idea one has to contemplate. That's the real McCoy."

"If what you have just said is right," Peter said, "I need to contemplate what you said in order to get it. Even so, I think I'll start using discursive reasoning anyway, albeit only because I'm not sure what contemplation is. Incidentally, I'm not even sure I know what you mean by 'discursive reasoning.'"

"Discursive reasoning is just the ordinary thinking we do all the time. Right now, as you think about what I've just said, you are involved in discursive reasoning."

"All right. And contemplation?"

"Ah, that's different. Maybe you should let me give you a kind of a formal definition of both. In discursive reasoning you put yourself, in relation to what you think about, in the subject/object mode. You, the thinker, see what you think about as a mental object in front of your mind's eye. You work with this mental object in this mode of separation between you and it. In contemplation, however, there is no separation between the thinker and that which he or she contemplates. They are one. If you start from the subject/object mode you experience the transition to contemplation as a kind of merging."

Julie was fully prepared for Peter's response: "That's a bit over my head, Julie. Give me an example."

"Remember Heisenberg's experience in Prunn Castle? He was listening to different speakers suggesting different kinds of order. He felt that 'all these types of order were partial, mere fragments that had split off from the central order,' but, try as he might, he 'was quite unable to find a way to the center through the thicket of conflicting opinions.' At this stage he was engaged in discursive reasoning. And then, something happened. A young violinist started playing Bach's Chaconne, and 'all at once, with utter certainty I had found my link with the center.' As I understand it, this was a moment of contemplation."

We sat quietly for a while, then Julie addressed both Peter and me: "I have noticed that discursive reasoning, while easily describable, brings no certainty. Thinking discursively about some issue, I often reach a conclusion and later change my mind. On the other hand, contemplation is not describable at all, and yet, as Heisenberg said, it comes with utter certainly. Why do you think that is?"

We both felt that Julie was getting at an important point, but neither of us could answer her question.

"Because," Julie explained, "as Plato said, discursive reasoning is superficial. All it can lead to are beliefs. Through contemplation one comes to the truth. When one comes to the truth, one knows it, with certainty."

Peter sank into himself, deep in thought. All of a sudden he exclaimed: "So that's what Schrödinger meant!" He was very excited. "Julie, remember the ending of Schrödinger's objectivation article? 'Subject and object are only one. The barrier between them cannot be said to have broken down as a result of recent experience in the physical sciences, for this barrier does not exist.'[14] I puzzled over this statement ever since we discussed that article. Now, in view of what you said, I get it. In contemplation, one is in touch with the truth. There is this merging you spoke about. One is in the Oneness of it all, so to speak. There is no subject and no object. It is only when one's mind reverts to the discursive mode that one objectivizes, that there is a bifurcation into the duality of subject and object. Schrödinger must have experienced what he wrote, through contemplation. It's all beginning to fit together."

Julie nodded approvingly, while Peter's active mind kept racing. "But, you know," he added, "something about this contemplation is still bothering me."

"Yes?"

"Heisenberg's experience was sudden. Doesn't contemplation take time?"

"Well, contemplation comes in many shapes and forms. It comes at different levels of depth and of purity and, yes, in different durations. Plato himself must have contemplated a long time to come up with something like the *Timaeus*. His teacher, Socrates, must have been quite a contemplator too. In the beginning of another dialogue, the *Symposium*, Plato describes how Socrates starts, all of a sudden, to contemplate in duration: Aristodemus speaks about walking with Socrates to Agathon's house, to have dinner there, 'and as we went along,' Aristodemus says, 'Socrates fell into a fit of abstraction and began to lag behind.'[15] 'A fit of abstraction,' for Socrates, does not sound like discursive reasoning to me. He was in the habit of practicing discursive reasoning with his followers all day long, not being 'abstracted' in the least. And if you want a more recent 'fit of abstraction,' here's one I just came across." She pulled a thick book from her bag. "This is a biography of Niels Bohr, written by Abraham Pais. Pais mentions Bohr's friend and colleague James Franck, who tells about his discussions with Bohr. Here's what he says." Julie read:

Sometimes he [Bohr] was sitting there almost like an idiot. His face became empty, his limbs were hanging down, and you would not know this man would even see. You would think that he must be an idiot. There was absolutely no degree of life. Then suddenly one would see that a glow went up in him, and a spark came, and then he said, "Now I know."[16]

All three of us became quiet. In this quiet we could hear the sounds of the surging waves with great crispness and clarity. Julie was moving her arm slowly, enjoying the sensation of her hand touching the sand, and as she did, out of nowhere, William Blake's lines came to her:

> To see a World in a Grain of Sand,
> And a Heaven in a Wild Flower,
> Hold Infinity in the palm of your hand,
> And Eternity in an hour.

11. "Nature Makes a Choice"

Continuing our exploration of the process of collapse, we raise and respond to the following questions: Why does a collapse occur? When does it occur? What brings it about? Paul Dirac's response to the last question is "Nature makes a choice." The idea that nature makes choices is reminiscent of Plotinus's idea that nature contemplates. The suggestion that during a collapse nature makes a choice indicates that in the context of the collapse objectivation fails: Nature cannot be treated merely as a collection of objects.

~

Nature, which they say is without imagination and without reason, has contemplation in itself, and produces what it produces by the contemplation which it "does not have."

—Plotinus

1. Keeping It Simple

In the previous chapter we became acquainted with the concept of the collapse of quantum states. We discussed the structure of the collapse, as well as its atemporal character. We are ready now to go deeper, and address the following three questions: First, what is the function of the collapse; why does it occur? Second, when does it occur? Third, what brings it about?

These questions have been the subject of intense research since the 1920s. The main proposed solutions are summarized in Appendix 3. As the discussion there shows, all of them are seriously flawed. The present chapter explains my own responses to these questions. These responses are based on a conversation I had with Paul Dirac back in 1976 (see Section 2).

Why does the collapse occur? Consider, once again, an electron moving toward a TV screen. Before it impinges on the screen it is a set of potentialities, i.e., a collection of all the possibilities for hitting each and every place on the screen. In other words, before impingement, the quantum state of the electron is a *superposition* of all the states that correspond to all the possibilities for hitting each and every place on the screen. All of these possibilities are somehow added together, or "superimposed." The quantum state includes them all, and then, in the atemporal process of collapse, one of them is singled out.

To understand why the collapse occurs, consider the alternative: What would it be like to live in a universe without collapse?

Let us just try to imagine such a universe. As our electron is moving toward the TV screen, we think of it as a superposition of all the states that correspond to all the possibilities for hitting each and every place on the screen. In this imaginary universe no choice as to which location will be hit is being made.

To be specific, by "location" we really mean "an atom." Obviously, the TV screen is made of atoms, and in *our* world, a world *with* collapse, the electron ends up interacting with just one among the trillions and trillions of atoms that make up the screen. We are considering, however, a universe without collapse. What would happen there?

To describe such a universe we can't help stretching the language, precisely because this imaginary universe is far from the universe of our experiences, and our language was never designed for its description. Please bear this in mind as we proceed.

To begin, a universe without collapse is a universe of potentialities rather than actual events, because the transition from the potential to the actual never takes place. Suppose, however, that potentialities could become actual without disturbing the superpositions. In other words, suppose that in every measurement the whole superposition, rather than one of the potentialities that make it up, would become actual. What then? Let us see.

The electron has hit the screen, but no atom has been singled out. In a sense, then, the electron has hit all the atoms. In a universe with collapse a particular atom is singled out. In a universe without collapse however, the superposition remains a superposition. Every atom on the screen is being hit "to a certain extent."

What next? In our universe, a universe with collapse, the hitting of one atom starts a series of processes that end with a visible flash occurring at the location of this atom. In a world without collapse, there will be trillions and trillions of flashes, flashes everywhere—each flash existing, however, "only to a certain extent," because each flash is still just an element in a superposition of flashes. Each of these flashes is, "to a certain extent," producing myriad photons, which affect the physical world around the screen in immensely complicated ways. The eye of the viewer, for example, will receive "to a certain extent" a myriad of photons not from one location on the screen but from all locations. The same applies to the atoms in the walls and in the furniture; they too will be affected by an immensely complex superposition of photons coming from everywhere on the screen.

In a universe without collapse, then, every interaction leads to a tremendous increase in complexity, and there is no process that works in the reverse direction, toward simplification.

Let us now come back to our own universe. We suggest that the answer to our first question, "Why does a collapse occur?" is this: *The function of the collapse is to simplify.* If there were no collapse, each and every interaction would make the universe immensely more complicated, and this trend toward increasing complexity would continue unchecked. In the real world, however, there is a process that reverses this trend—the collapse! Because of collapse, the level of complexity increases and decreases as quantum systems switch back and forth between the potential and the actual. An electron that is moving toward a screen is in a state of superposition of possibilities. The impingement upon one particular atom on the screen means that the superposition has "collapsed," the electron has become actual, and its quantum state has become simpler: The electron is interacting with a single atom. As a result of this interaction, and the processes that follow, photons are emitted. Each photon is a new field of potentialities—a superposition of possibilities as to the choice of the atom it (the photon) will hit. The situation is complex again. Now there are new collapses, each photon interacts with one particular atom, things are simpler again, and so on.

Collapses occur because they are essential ingredients in the interplay of opening up and closing down. As the actual becomes potential, there is an opening up into a superposition of many coexisting possibilities. As the potential becomes actual, there is a closing down: Most potentialities are denied actualization, as one of them is singled out. The chosen one becomes an actual event, only to change immediately into a superposition of possibilities for the next collapse. And so the dance goes on.

2. The Timing

When does the collapse occur?

In 1976 I attended a very small conference at Florida State University in Tallahassee. The conference was devoted to "Dirac cosmologies"—certain ideas about cosmology that Paul Dirac was exploring at the time. Dirac initiated the conference, and his purpose was to discuss his ideas with a small group of researchers in cosmology.

Most of our discussions were indeed devoted to cosmology. One sunny afternoon, however, we all headed to Wakulla Springs to take a boat ride and see the alligators. As we were driving out I had a chance to talk to Dirac and ask him a question that had nothing to do with cosmology. This question had been on my mind for years (as it was, and is, on the minds of most researchers of the foundations of quantum mechanics). I asked him, "How does a collapse occur?" His response was "Nature makes a choice." I then asked, "When does nature make a choice?" and he said, "When there is no longer a possibility of interference." I was impressed. I felt intuitively, on the spot, that what Dirac had said was the key to a novel and correct understanding of the issue of collapse.

We will come back to the first statement, "Nature makes a choice," in Section 4. The second response, "[The collapse occurs] when there is no longer a possibility of interference," is the answer to the question of timing. The rest of the present section is devoted to an explanation of this response.

We have come across the term "interference" before. In Section 2 of Chapter 3 we discussed the principle of superposition, and gave, as an example, the interference of water waves moving across the surface of a lake. When two waves meet in such a way that they rise and fall together, they reinforce each other and are said to interfere constructively. When they meet in such a way that when one wave moves up the other moves down, they destroy each other, and are said to interfere destructively. These are the extreme cases. In other, in-between situations they reinforce (or destroy) each other only partially; their interference is neither totally constructive nor totally destructive.

When two, or three, or four waves cross each other's paths, a beautiful interference pattern is seen on the surface of the water. It is even possible for the waves to separate and recombine; their coming together and interfering does not do away with the possibility of separation and subsequent recreation of the original pattern. But what if it starts raining, and the wave pattern becomes subject to millions of disturbances? The original interference pattern is lost forever—even when the rain stops, it will not reappear.

The principle of superposition does not apply to water waves alone. It applies to all waves, including Schrödinger's waves of potentialities. These waves too interfere, reinforcing or weakening each other, as the case may be. Here too, a few interfering waves can separate and recombine in clear patterns; but when a wave pattern breaks up into a zillion little waves, a recombination is no longer possible, and a collapse occurs.

According to Dirac, the collapse occurs when the different possibilities can no longer interfere with each other to create a coherent pattern. As our example shows, this issue of interference is closely related to the issue of complexity. When the relationships among the different possibilities are so complex as to be chaotic, a coherent interference pattern is no longer possible, and a collapse takes place.

The impingement of an electron upon a TV screen is an example of such a break-up. Unless a collapse occurs, the resulting wave pattern is hopelessly complex, and its many elements can never recombine into a clear, distinct shape. On the other hand, when a double-slit interference experiment is carried out, the two waves that propagate from the two slits are perfectly capable of recombining. The recombination gives rise to the distinct pattern depicted in Figure 4.1. The presence of the barrier with two openings does not lead to a collapse.

The processes that generate enough complexity to make coherent interference impossible, processes that lead to "decoherence," have been a lively subject of research in the past two decades. Calculations show that unless special precautions are taken in the construction of an experimental set-up, decoherence occurs very quickly. Processes in nature are characterized by a quick succession of decoherence and collapse. It is not easy to persuade nature not to make a choice!

3. Can Choices Be Avoided?

Let us clarify the subtleties of superpositions and collapse with the help of an experiment. Figure 11.1 shows an experimental set-up involving ordinary mirrors as well as half-silvered mirrors. (A half-silvered mirror is a piece of glass coated with a layer of silver so thin that it reflects half the light and transmits half the light, rendering glass half transparent and half reflective.)

The set-up depicted in Figure 11.1 is a stage on which a drama of separation and recombination of potentialities takes place: A beam of light, arriving from the left, is split in two by the half-silvered mirror. The reflected half of the beam moves upward along path a, and the transmitted half moves right along path b. The reflected half of the beam is redirected by the two ordinary mirrors at the top of the drawing, to meet the transmitted half. The meeting occurs as both half-beams arrive at another half-silvered mirror, the one on the right. Now two more split-tings take place, as each half-beam is split in two by this second half-silvered mir-ror. The half-beam that followed path a splits into (1) a quarter-beam reflected toward the right and (2) a quarter-beam that passes right through and continues downward. The half-beam that followed path b is split into (3) a quarter-beam that continues straight and is moving toward the right and (4) a quarter-beam that is reflected downward.

Consider quarter-beams 1 and 3. They are now traveling together and follow-ing the same path; hence, according to the principle of superposition, they inter-fere. Their interference may be constructive or destructive, but let's not worry about that. The issue we need to consider now is the transition from the classical to the quantum domain.

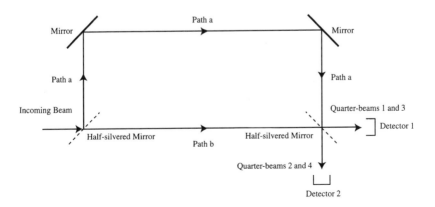

Figure 11.1 An experimental set-up in which a beam of light is separated into two beams, which subsequently recombine.

The set-up we described is a classical one. Nineteenth-century physicists exper-imented with such set-ups and verified the expected interference patterns. What happens, however, when we turn the brightness of the incoming beam way down? Let us turn the brightness down so much that the incoming beam becomes a flow of individual photons: One photon arrives, goes the whole way, and leaves the apparatus; then another photon arrives, and then another. . . . We can no longer say that the half-silvered mirrors are splitting beams of light. We have to consider what happens to each individual photon as it arrives at these mirrors.

What happens, then, when a photon arrives at the half-silvered mirror on the left? Will it be reflected or transmitted? We may be tempted to say: The chances of reflection and transmission are equal; hence half the photons will be transmit-ted, and half reflected. As to this particular photon we are considering at this moment, it may be reflected, or it may pass through; there is no way to tell.

This answer is in the spirit of quantum mechanics, but it is wrong. It is wrong because *it is based on the assumption that a collapse takes place, while, in fact, no col-lapse occurs.* Look at it carefully: The incoming photon, upon hitting the half-sil-vered mirror, has two possibilities open to it, reflection and transmission. Nothing, however, forces it to make a choice. It will be, therefore, in a state of superposi-tion of the two potentialities, the potentiality for going along path a, and the poten-tiality for going along path b. This superposition does not collapse; a superposi-tion of only two states is not complex enough to warrant a collapse.

What happens at the half-silvered mirror on the right? Again there is no rea-son for collapse. The state of the photon is now a superposition of the four possi-bilities 1, 2, 3 and 4. The superposition of 1 and 3 is particularly interesting. These two possibilities refer to the same path. In classical physics one speaks of such a superposition as an interference between two steady beams of light, traveling together. In quantum physics one speaks instead of the interference between two potentialities referring to the same photon. The photon, being nothing more than a field of potentialities, interferes with itself! Its potentiality 1 interferes with its potentiality 3!

Can the interference between potentialities 1 and 3 be verified experimentally? If one observes only one photon, the answer is no. Eventually, when a collapse finally occurs, this one photon will be detected as one dot, flashing somewhere; the detection of one flash cannot lead to any conclusion about interference. The interference between potentialities 1 and 3 can be verified, however, by looking at the pattern which appears when the experiment is repeated many times, with many photons. This is analogous to the double-slit experiment. When our exper-iment is repeated many times, the rate of arrival of photons at the detectors that have been placed to the right of the apparatus is determined by the interference between potentialities 1 and 3. If instead of an interference there were a collapse, and a choice between path a and path b had been made for each photon, the rate at which photons arrive at the two detectors would have been quite different.

Our set-up is such that photons pass through it without collapsing. If we change

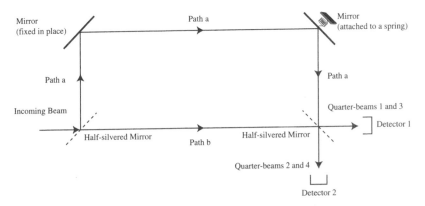

Figure 11.2   A variant of the experimental set-up depicted in Figure 11.1: The mirror at the top right is now mounted on a light spring.

it only slightly, however, a collapse will occur every time. In the set-up depicted in Figure 11.1 all the mirrors are fixed in place. Consider now a variant of this set-up: Let one of the fully silvered mirrors, the one on the top right, be mounted on a light spring (Figure 11.2). This spring is so light that it will start vibrating whenever anything, even something as tiny as a single photon, impinges upon it.

The presence of this light spring is enough to destroy the possibility of interference between potentialities 1 and 3; it is enough to bring about a collapse. The spring forces a decision: If a photon goes through path a, the mirror will start vibrating; if it goes through path b, the mirror will stay still. The mirror cannot be in a state of superposition of vibrating and not vibrating. Such a state requires extremely complicated superpositions of the states of the trillions of atoms that make up the mirror; and, as far as the photon is concerned, the vibration of the mirror blurs its quantum state to such an extent that a coherent interference between the potentialities of traveling along path a and path b is no longer possible.

Let us compare the two set-ups. In Figure 11.1 the photon is not "choosing" a path. It gets through the apparatus as a superposition of potentiality waves traveling through paths a and b. In Figure 11.2 the photon is "choosing" a path. It gets through the apparatus traveling either through path a or through path b; the choice between a and b is made at random for each photon.

4. Who Done It?

What brings the collapse about?

A choice is about to be made—an electron is about to impinge upon a TV screen, and it can hit any location on the screen. Quantum mechanics provides the distribution of probabilities, but only one location will be hit. How is the choice being made, and what is the agent that makes it?

Dirac's statement "Nature makes a choice" opens up an unexpected line of inquiry. What was Dirac's notion of "nature" as an entity ("entity" is the wrong word, but our language does not seem to have a right one) that can make a choice? The issue goes even deeper: Nature makes a choice *and acts on it.* The choice results in an act of creation; the transition from the potential to the actual is a creation of an actual event! Unfortunately, I did not explore these issues with Dirac when I had the chance. Over the years, however, these questions continued to interest me deeply. I don't know how Dirac thought about nature, but the new ways of thinking about nature that his statement opened up for me are crucial, I believe, for the transition to the new paradigm—a paradigm that is broadly indicated by the characteristics of the collapse of quantum states.

In Chapters 8 and 9 we saw that the tenet of realism is suspect. Both Whitehead's fallacy of misplaced concreteness and Schrödinger's principle of objectivation expose realism as a mere abstraction. Just like the assumption of the flat earth, a realistic mode of thinking may lead to correct conclusions in certain circumstances; it is a mistake, however, to consider realism a basic feature of the way things are.

As we will see, the question of the collapse of quantum states, and, in particular, Dirac's statement "Nature makes a choice," open up novel ways of thinking about reality, ways which provide independent support for Whitehead and Schrödinger's realizations that the assumption of realism is untenable. The search for the new, emergent paradigm, a paradigm that includes an answer to the question "What brings the collapse about?" will occupy us for the remainder of this book. And we will be helped by the wisdom of a great sage.

Plotinus, who lived in the third century A.D., was the last of the great Platonic philosophers of antiquity. Section III.8 of his *Enneads* is largely devoted to an exposition of a key element of his thought, the idea that *nature contemplates, and creates by contemplation.* Appreciating how strange the idea would appear to minds not yet prepared for it, Plotinus introduces it cautiously.

> Playing at first, before we handle our subject seriously, if we were to say that all things desire contemplation and look to this end, not only rational but also irrational animals, and the nature in plants and the earth which engenders them, and that all gain it, each one having it according to the nature of each, but that different ones gain contemplation in different ways, some truly, while others get an imitation and image thereof—would anyone sustain so paradoxical a statement?[1]

The sequel leaves no doubt that this "playing" is, for Plotinus, of the utmost seriousness. The idea that nature, as well as most other elements of reality, contemplates is central to his cosmology.

Nature contemplates, and creates by contemplation. Can this be the key to a new world-view—one that integrates ancient wisdom and contemporary knowledge?

5. Quantum Mechanics and the New Paradigm

Since quantum mechanics is based on the principle of objectivation it cannot provide us with a new world-view. Yet we have gleaned, in the previous section, a clear indication about the nature of reality from quantum mechanics. How did we manage to do that?

Quantum mechanics is not your run-of-the-mill theory. It is a theory that harbors a deep mystery: the collapse of quantum states. It is not an accident that the indication about the nature of reality was gleaned from the discussion of the collapse. For the past six or seven decades the collapse has been a source of difficulty for researchers of quantum mechanics. In spite of a great deal of hard work (see Appendix 3), no one has yet succeeded in discovering a mechanism for the collapse. No one has managed to explain *how* nature chooses one possibility rather than another.

This persistent difficulty led me to suspect that the mechanism of collapse has not been unraveled because there is no mechanism. And there is no mechanism because, as Whitehead and others have suggested, nature is an organism whose functioning cannot be reduced to a set of mechanisms. *This persistent difficulty shows the limit of applicability of the principle of objectivation.* The assumption that nature can be considered an objectivized entity is an abstraction. Every abstraction has a limited domain of validity. The atemporal process of collapse is outside of this domain. The collapse cannot be unraveled as a part of the formalism of the quantum theory because nature is not *really* objectivizable!

Quantum mechanics has given us an important indication about the direction of the new paradigm. Because it shows us precisely where the abstraction of the principle of objectivation fails, we can learn how this principle can be transcended. This transcendence is what we turn to next.

12. Nature Alive

Our analysis of the collapse of quantum states indicated that the principle of objectivation has to be transcended. The universe and its constituents are alive! This view is supported by eminent scientists' reports of direct experiences of the aliveness of putatively inanimate objects.

~

It is preposterous to make the world soulless when we, who contain a part of the body of the All, have a soul.

—Plotinus

1. Beyond Objectivation

Nature contemplates and creates by contemplation. What a strange idea! We are in search of a new world-view, but do we have to go *this* far? Let us keep this question in abeyance for the time being. We will come back to it in Chapter 17; the intervening chapters will prepare the ground.

The present chapter is devoted to the realization that nature is not really objectivizable. The difficulties that arise when we try to get a full grip on the atemporal process of collapse indicate the limits of the applicability of the principle of objectivation. An unobjectivizable aspect of nature is revealed. What does this mean?

Evidently, when the entities in nature are treated as objects, not all of their characteristics show up. There is something about them that cannot be objectivized. "An object" is how an entity appears *to us*. Its appearance as an object is its *public* character. If an entity cannot fully reveal itself as an object, it must have another *private* aspect that is in and for the entity itself. In other words, *it must be alive*.

What is the meaning of the statement "An entity is alive"?

J. M. Burgers was the first scientist to notice that Whitehead's Philosophy provides quantum mechanics with a sound metaphysical basis (see Chapter 15). In his book, *Experience and Conceptual Activity*, subtitled *A Philosophical Essay Based Upon the Writings of A. N. Whitehead*, he writes, "...life cannot be completely described or analyzed in terms which have fixed meanings in the vocabulary of physics and mathematics. There are essential features of life that do not have an unambiguous relationship to states of matter or of fields that can be objectively characterized. Their explanation requires reference to subjective features."[1] One

135

of these subjective features is the creativity, which forms the basis of Burgers' definition of life: "We consider life as a result of the creativity which we have assumed [with Whitehead] as a fundamental feature of the universe. In order to distinguish the notion of life from creativity in general, we say that life is a coordinated and sustained form of creativity, the coordination and persistence serving to enhance the extent of the creativity. As creativity involves an amount of freedom, life can be seen as the *persistent struggle to retain freedom*."[2] [italics in the original]

An alternative definition, one that expresses life *from the point of view of the living entities* is this: *An entity is alive if, in principle, it is capable of experiencing*. The statement that butterflies, plants and bacteria *experience* may seem strange. As we will see in Chapter 15, Section 2 we are using the word "experience" in a generalized sense, a sense that makes the statement plausible and complements Burgers' definition.

2. The Wisdom of the Sphinx

Sometimes the relationship between living entities is not as simple as it seems; consider the philosopher P. D. Ouspensky. When he was in Egypt, he was sitting every day for many hours in front of the Sphinx. There was no question in his mind that the Sphinx was alive. The passage of time for the Sphinx, however, was so different than for himself, that he felt that the Sphinx could not see him, "and not because I was too small in comparison with it or too insignificant in comparison with the profundity of wisdom it contained and guarded—the sense of annihilation and the terror of vanishing came from feeling myself too transient for the Sphinx to be able to notice me."[3]

Let us finally address the following question: We claimed that everything is alive. Well, how about a corpse? Is it alive too?

From the point of view of one's own experience, the difference between a person and a corpse is that one can have moments of real relationship with a person, and not with a corpse. And if a direct relationship is impossible, e.g. between Ouspensky and the Sphinx, or between a man and an ant, intermediate living things may provide the missing link.

3. "Ye Presences of Nature"

Ye Presences of Nature in the sky
And on the earth! Ye visions of the hills!
And Souls of lonely places! can I think
A vulgar hope was yours when ye employed
Such ministry, when ye through many a year
Haunting me thus among my boyish sports,

On caves and trees, upon the woods and hills,
Impressed upon all forms the characters
Of danger or desire; and thus did make
The surface of the universal earth,
With triumph and delight, with hope and fear,
Work like a sea?

In these beautiful lines from the poem The Prelude, quoted by Whitehead,[4] Wordsworth expresses his immediate experience of the aliveness of nature. Do these lines express a true fact, or should they be enjoyed as mere poetic imagination?

I believe that the proposition "Nature is alive" is an obvious truth that we take pains to avoid. As we go about our business, both in science and in everyday life, we just ignore it, adhering, as we do, to the obsolete Newtonian framework and to reductionistic methods of analysis and explanation. Yet, the more sensitive and perceptive among us cannot help but express the conviction that putatively inanimate things are alive. We are accustomed to encountering such expressions in poetry. A close reading of Wordsworth's lines shows that as far as he is concerned, the presences he writes about are real. And poetry is by no means the only context in which the aliveness of the so-called inanimate is stated as a matter of fact. Such statements do appear, out of the blue, and are voiced by the most unlikely people on quite unexpected occasions.

4. From Mitchell's Harmonious Planet to Jensen's Angry Sculptures

Consider the astronaut Edgar Mitchell. Being an astronaut, he saw the whole earth as it was floating weightless in outer space. He had a very strong experience, a kind of a conversion. It was triggered by the fact that he *saw* the whole earth as alive. "There seems to be more to the universe," he said, "than random, chaotic, purposeless movement of a collection of molecular particles. On the return trip home, gazing through 240,000 miles of space toward the stars and the planet from which I had come, I *suddenly* experienced the universe as intelligent, loving, harmonious."[5]

I italicized the word "suddenly" for a good reason. Clearly, at the moment he referred to Mitchell went through a major paradigm shift. Such shifts are always cha-racterized by being sudden, unexpected.

Next, consider the late Dr. Lewis Thomas, an establishment figure *par excellence*. During a long and distinguished career he had been president of the Memorial Sloan-Kettering Cancer Center, dean of the Schools of Medicine of New York University and Yale, and a professor of pediatrics, medicine, pathology, and biology. Thomas' impeccable credentials within the establishment gave him the freedom to critically examine cherished assumptions. An address he gave in New York in 1984 contains the following passage:

The most beautiful object I have ever seen in a photograph, in all my life. is the planet earth seen from the distance of the moon, hanging there in space and obviously alive. Although it seems at first glance to be made of innumerable separate species of living things, on closer examination every one of its working parts, including us, is interdependently connected to all the other working parts. It is, to put it one way, the only truly closed ecosystem any of us knows about. To put it another way, it is an organism and it came into being about four and one half billion years ago. It came alive, with a life of its own, I shall guess 3.7 billion years ago today.[6]

Both Mitchell and Thomas's quotations are significant not only because they express the idea that the earth is alive, but because they express it as intuitively obvious. It is this immediate impact of the aliveness of the earth that flies in the face of the generally held scientific presumption that the earth is nothing more than a chunk of inanimate matter. As Whitehead puts it, "*If one is in touch with immediate feelings and intuitions one can see how strained and paradoxical is the view of nature which modern science imposes on our thoughts* [italics added]."

A different testimony is provided by a physicist of impeccable credentials— Erwin Schrödinger. Schrödinger points out that scientific descriptions of phenomena involving life can be misleading. In his article "The Principle of Objectivation," which we discussed in Chapter 9, Schrödinger objects to the standard scientific view that the eye is a purely receptive organ, an organ that allows light to come in and does not send anything out. He writes:

I wonder has it ever been noted that the eye is the only sense organ whose purely receptive character we fail to recognize in naïve thought. Reversing the actual state of affairs, we are much more inclined to think of "rays of vision," issuing from the eye, than of the "rays of light" that hit the eyes from outside. You quite frequently find such a "ray of vision" represented in a drawing in a comic paper, or even in some older schematic sketch intended to illustrate an optic instrument or law, a dotted line emerging from the eye and pointing to the object, the direction being indicated by an arrow-head at the far end.[7]

5. Knud Jensen's Angry Sculptures

The Louisiana Museum in Humlebaek, Denmark, owes its name to three ladies called Louise; it has nothing to do with the state of Louisiana. In 1981 Lawrence Weschler, staff writer for *The New Yorker*, published an article about the museum and about its founder and curator, Knud Jensen.[8] Like Dr. Thomas, Jensen is no idle dreamer. He managed to create and direct, with little outside help, a museum with an impressive collection as well as a unique atmosphere. The museum is unique in providing what Weschler calls "an unparalleled" sense of human provenance of art." At the Louisiana Museum, writes Weschler, "people milled in the late-afternoon sun, sat on picnic blankets, slept on the lawn, or cuddled in the shade, more like neighbors than visitors. There was a great sense of relaxation, of familiarity."

Jensen is a visionary (how else could he create such an unusual museum?), but he is also a man of the world, focussed, pragmatic, and effective. This practical man has no doubt that works of art are alive, and he treats them with the appropriate respect. For example, throughout a wide lawn, just outside the museum building, there are only three pieces of sculpture. When Weschler asked him why, Jensen replied:

It's strange—this whole wide lawn, and all it can take is three pieces. I've noticed that you can put in five, seven, ten pieces of sculpture. It's as if the pieces were slightly tamed by the cage. But against a natural backdrop you can seldom have more than three, no matter how large the space. You know the work of Konrad Lorenz, the great ethnologist—his writings on animal behavior and territoriality. That's why the sculptures at Louisiana are spaced as they are: Lorenz had shown that all living things stake out their natural territory, require their breathing space. Sculptures, like all great art, are, of course, living things. I have tried to move another piece onto the lawn here, but the others got very angry; they withheld their luster as if in resentment.

Is Jensen's statement a metaphor, or is it literally true? Being a product of contemporary Western culture, I find it difficult to accept Jensen's explanation at face value. But Jensen speaks about the aliveness of his pieces of sculpture as a matter of fact. It is obvious to this practical, competent man that they are simply and truly alive. How could that be? After all, these are just pieces of stone, aren't they? As I experience this clash between my intuition and customary ways of thinking, I wonder: Should we follow our own experience, or shall we let ourselves be convinced that the emperor is clothed?

A few years ago I visited the Egyptian collection at the Boston Museum of Fine Arts. After strolling for a few moments, I stopped in front of a piece of sculpture forty-five centuries old, the bust of Prince Ankh-haf. As I looked at it, it became obvious to me that sculpture was alive and *that we were communicating*. The communication was entirely non-verbal, and yet, when I left, I felt strangely uplifted, as if I had emerged from an extraordinary meeting.

I could not deny my experience, and yet I could not fit it into the science-based paradigm, which, under ordinary circumstances, provides me with a framework of explanation for the events of my life. When I remembered what a paradigm is, however, the acute discomfort began to lift. A paradigm is nothing more than an abstraction adopted by a society at a given period in its history. Every abstraction has a restricted domain of validity. Unfortunately, members of the society implicitly accept the abstraction as absolute truth. One is conditioned by one's culture to explain away or just ignore experiences that indicate the existence of domains where the paradigm is invalid. Once in a while, however, this conditioning breaks down, and one has *a new experience*—an experience that does not fit the paradigm. Such experience can be invaluable as indications of the limitations of the paradigm.

6. Science and the Life of "Inanimate Things"

It may be shocking to realize that *according to the Newtonian world-view one's aliveness has nothing to do with one's actions:* The universe runs like clockwork; the present and the future are completely determined by the past. It follows that our decisions and intentions can have no bearing on our actions. After all, the body is just a collection of atoms, whose movements are totally predetermined. It follows that I have as much control over the movements of my hands as I have over the movements of the moon: Both movements are the inescapable result of the configuration of matter in the universe in the remote past.

The idea of a clockwork universe brought about a kind of schizophrenia in the culture. The functioning of science was firmly based on the Newtonian paradigm of a clockwork universe, and yet, in one's private world, as well as the world of social intercourse, decisions and intentions were not considered illusory. People were held responsible for their actions.

Within a Newtonian framework one has no choice but to think of the events in the physical world as predetermined movements of a lifeless machine. Now that the Newtonian world-view has collapsed, however, this belief in the lifelessness of natural processes makes no sense. Within physics, the Newtonian paradigm is no longer a valid framework for the conceptual understanding of reality: Its notions of space and time were superseded by Einstein's theories of Special and General Relativity; its notion of matter was superseded by quantum mechanics. But since science has failed to provide us with an alternative paradigm, and thinking abhors a vacuum, the Newtonian paradigm lives on. Hence our uneasy reaction to Dr. Thomas's statement about the earth, or to Jensen's description of his resentful pieces of sculpture.

This belief in the lifelessness of the universe and its constituents (with very few exceptions) is reinforced, first, by the current scientific mentality, which is wholeheartedly devoted to the reductionistic analysis of parts, and, second, by the tacit adoption of the principle of objectivation.

Let us take the second point first. Science cannot attest to the proposition that the universe is alive, because the uncritical denial of this proposition is at the heart of the abstraction that science is based on, i.e., the principle of objectivation. The denial is uncritical, because the principle is adopted implicitly. As Schrödinger puts it, our unawareness of the fact that a moderately satisfying picture of the world has only been reached at the high price of taking ourselves out of the picture ...[leads to] the astonishment at finding our world-picture "colorless, cold, mute." Color and sound, hot and cold are our immediate sensations; small wonder that they are lacking in a world-model from which we have removed our own mental person.[9]

The reductionistic approach of science, i.e., the conviction that the understanding of the whole can be arrived at by an analysis of the parts, is another factor that goes against the recognition of the aliveness of things. To be sure, the reductionistic approach is useful and powerful. For example, without a thorough examination of the parts of living organisms, we could not have arrived at the discovery of the genetic code and at all the incredible possibilities that this discovery opens up. The reductionistic approach, however, has its limitations. There are essential aspects of the whole to which it is blind.

Bohr's framework of complementarity can clarify the issue. There are essential aspects of the whole that can only be grasped by the experience of the whole. The analysis of the parts and the *experience* of the whole are complementary; they cannot be carried out at the same time, but both are needed for a complete understanding. This point is beautifully illustrated by Arthur Eddington, an astronomer and astrophysicist who was active at the time of the birth of quantum mechanics. While not being himself one of its founding fathers, he followed this birth rather closely.

Eddington writes about the relationship between "symbolic knowledge" and "intimate knowledge." Although he does not use the term, he explains this relationship as one of complementarity:

For illustration let us consider Humor. I suppose that humor can be analyzed to some extent and the essential ingredients of the different kinds of wit classified. Suppose that we are offered an alleged joke. We subject it to scientific analysis as we would a chemical salt of a doubtful nature, and perhaps after careful consideration of all its aspects we are able to confirm that it really and truly is a joke. Logically, I suppose, our next procedure would be to laugh. But it may certainly be predicted that as a result of this scrutiny we shall have lost all inclination we may ever have had to laugh at it. It simply does not do to expose the inner workings of a joke. The classification concerns a symbolic knowledge of humor which preserves all the characteristics of a joke except its laughableness. The real appreciation must come spontaneously, not introspectively.[10]

As Eddington demonstrates so clearly, these complementary modes (the experience of the whole and the analysis of the parts) are mutually exclusive. We cannot examine the components of a joke and simultaneously experience it as a cause for laughter. Similarly, the aliveness of things does not show up through an analysis of their parts; it becomes evident through experiences of the whole.

Our discussion leads us to the following conclusion: The scientific claim that so-called inanimate entities are really lifeless is a statement about the scientific method and not about the entities. In reality there is nothing in the current scientific knowledge that disproves the proposition that putatively inanimate entities are alive. This proposition can only be verified or disproved through experiences and modes of knowledge that lie outside of the methodology of present-day science. It is entirely possible that in the future science will outgrow the principle of objectivation, as well as the reductionistic bias, and will come to encompass different kinds of experience within its ken. As long as it is based on the principle of objectivation and relies on the reductionistic mode, however, it is in no position to pass judgment on the aliveness of things.

This is an important point. When scientists scoff at the idea that the universe is alive, they express their views as private individuals, not as scientists. Claims that affirm or deny the proposition that the universe is alive in the name of science are simply invalid: If the universe is indeed alive, or "ensouled" as the ancient Greeks put it, this aliveness will not show up in a scientific context. Such claims merely demonstrate scientists' ignorance regarding the distinction between abstraction and concrete fact.

7. The Soul of the Universe

Our culture is addicted to science. We look up to science for answers to questions that are outside of its domain of knowledge. Once we realize that the scientific approach is as limited in scope as it is powerful within its scope, we will cease trying to draw water out of a dry well.

If present-day science is of no use to us in relation to the question of whether or not the universe is alive, where should we turn? Well, we saw in Sections 2–5 that if we turn to the intuitively obvious, to immediate experiences, the evidence is compelling. And because we are conditioned to discard such experiences in favor of the abstractions of science, their occurrence is all the more remarkable. Any yet, immediate experiences are not the sole witnesses to the aliveness of things. As soon as we turn away from present-day science, we discover that the idea that the universe and its constituents are alive is an integral part of almost all paradigms! Not only is it encountered in all non-Western cultures, it is a firm element in practically all schools of Greek science, the source of our own scientific tradition.

G. E. R. Lloyd examined the issue in some detail. He presents "the notion that the primary substance of things is in some sense instinct [i.e., profoundly imbued] with life, and the idea that the world as a whole is (or at least is like) a living organism" as a major theme in Greek philosophy. Anaximander "thought of the evolution of the world as the growth of a living being." Heraclitus calls "the world-order itself an ever-living fire which 'kindles in measures and goes out in measures.' Fire is not only eternal, but also, it seems, in some sense alive. That 'ever-living' is not merely a poetic equivalent for 'everlasting' is obvious if we reflect that Heraclitus believed that fire is the substance of which our own souls are composed....The Pythagoreans too, like Anaximander and perhaps also Anaximenes, assumed that the world as a whole is alive and described its origin and development in terms of those of a living organism."[11]

If the universe is alive, emotions may well have cosmological significance. Thus "following the lead of Parmenides, perhaps, Empedocles described the abstract cosmological principle which unites the four elements as Love (and explicitly referred to the works of Love among men to illustrate and exemplify the principle)."[12]

In the *Timaeus*, Plato's account of the creation of the universe, the physical universe is conceived not only as having a soul but also as being contained within it:

> Now when the creator had framed the soul according to his will, he formed within her the corporeal universe, and brought the two together and united them center to center. The soul, interfused everywhere from the center to the circumference of heaven, of which also she is the external envelopment, herself turning in herself, began a divine beginning of never-ceasing and rational life enduring throughout all time.[13]

In spite of great differences among their systems, most Greek philosophers held the position that the universe is alive because its aliveness was, for them, intuitively obvious. This position held throughout antiquity. For Plotinus, in the third century A.D., the idea that the world is not alive was totally unreasonable. He agrees with Plato that "it is preposterous to make the world soulless when we, who contain a part of the body of the All, have a soul."[14]

Lest we be left with the impression that this idea that the universe is alive is somehow linked to "primitive" or "naive" traditions, traditions we have outgrown through the discoveries of science, we have Whitehead's monumental achievement to contend with. Whitehead's "process philosophy" is based on a critical evaluation of previous philosophical systems, on the upheaval in physics and the discoveries in mathematics at the beginning of the twentieth century, and also on a perceptive analysis of the evidence of immediate experience, his own and others'. Having had a distinguished career as a mathematician before he turned to philosophy, he used his powerful intellect to create formulations that are as logically impeccable as they are intuitively perceptive. And Whitehead's philosophical system (see Chapter 15) is firmly based on the proposition that the universe is an organism and that its basic constituents are "throbs of experience."

8. Being Touched by the Soul of the Universe

In Section 3 of Chapter 10 we quoted Heisenberg's description of his experience of making contact with "the central order." Years later, when his friend Wolfgang Pauli asked him if he believed in a personal God, this experience informed his response:

> May I rephrase your question? I myself should prefer the following formulation: Can you, or anyone else, reach the central order of things or events, whose existence seems beyond doubt, as directly as you can reach the soul of another human being? I am using the term "soul" quite deliberately so as not to be misunderstood. If you put your question like that, I would say yes.[15]

Erwin Schrödinger had the experience of finding the soul of the universe within himself, as his own ultimate identity. He expressed his finding as follows:

> Inconceivable as it seems to ordinary reason, you—and all other conscious being as such—are all in all. Hence this life of yours which you are living is not merely a piece of the entire existence, but is, in a certain sense, the *whole*; only this whole is not so constituted that it can be surveyed in one single glance. This, as we know, is what the Brahmins express in the sacred, mystic formula which is yet really so simple and so clear: *Tat twan asi*, this is you. Or again, in such words as "I am in the east and in the west, I am above and below, *I am this whole world*" [italics in original].[16]

Albert Einstein responded in a letter to a sixth grader's question about religion as follows:

> Every one who is seriously involved in the pursuit of science becomes convinced that a spirit is manifest in the laws of the Universe—a spirit vastly superior to that of man, and one in the face of which we with our modest powers must feel humble. In this way the pursuit of science leads to a religious feeling of a special sort, which is indeed quite different from the religiosity of someone more naive.[17]

These great scientists worked within the confines of the principle of objectivation, but through their own experiences they knew that there is more to reality than objects moving in space. In one guise or another, they were touched by the soul of the universe.

13. Flashes of Existence

We now continue to study the properties of elementary quantum events as they appear under scientific scrutiny. According to Newtonian physics, matter is made out of enduring objects called "atoms." By contrast, according to quantum mechanics, the basic units of the physical world are events that flash in and out of existence. Thus matter is made up of units that are discrete not only in space but also in time.

~

Those who are not shocked when they first come across quantum theory cannot possibly have understood it.

—Niels Bohr

1. Elementary Quantum Events: Now You See It, Now You Don't

If the universe is alive, what about its fundamental units, the happenings that show up as "elementary quantum events?" Are they alive as well? What are these units really?

This last question has two aspects. First, what is their nature *as perceived by us?* Second, what is their *real* nature, *independent of our perception?* While the first question seems fair enough, the second question seems preposterous to the point of being meaningless. How can we possibly know this real nature? How can we even know whether or not they have one?

In the present chapter we will address the first question. A critical evaluation of the meaning or meaninglessness of the second question will be taken up in Chapter 14. Following this evaluation we will answer the second question in Chapter 15. Let us now proceed to find out the characteristics of these fundamental units of nature that are revealed in scientific measurements.

The collapse of quantum states is an elementary act of creation. When conditions call for it, elementary quantum events appear out of an underlying field of potentialities. Unlike these underlying fields of potentialities, which do not exist in spacetime, these events are actualities. What are these events like?

First and foremost, elementary quantum events do not last long. Rather, they exist as flashes. *An elementary quantum event appears suddenly, and disappears almost as soon as it appears.* Let us consider, once again, an electron moving toward a TV screen. The expression "an electron moving toward a TV screen" does not

145

describe an actual process in spacetime. Rather, it says something about how the distribution of probabilities changes in time. The distribution keeps changing, and then, suddenly, there is a collapse. Instead of a field of potentialities there is an impingement upon a particular location on the screen. At that instant *an actual event is taking place in the physical world.* A short time later, however, it is potentialities all over again. To be sure, the atoms of the TV screen are affected; they precipitate the processes that ultimately lead to the appearance of a visible dot on the screen. But this appearance has to do with many other particles and many other processes of collapse. The electron that started it all by collapsing has disappeared from view. It has become, once again, a field of potentialities.

Let us change our perspective now and look at a collapse not from the point of view of the event but from the point of view of the potentialities. This can be described as follows: *A collapse is a sudden, discontinuous change in the otherwise smooth, wave-like evolution of potentialities.* According to Schrödinger's equation, as long as there is no collapse, the evolution of potentialities in time is smooth and continuous. The collapse introduces a sudden discontinuity. Before impingement, the possible places of impingement are spread all over the screen. When a collapse occurs there is, all of a sudden, no possibility of finding the electron anywhere except at the one location of the actual impingement. Once this discontinuous change has taken place, however, the field of potentialities starts evolving again, smoothly and continuously. The sudden impingement acts like a pebble dropped into a quiet pool of water: Ripples begin to spread. New potentiality waves begin propagating from the place of impingement smoothly and continuously.

This interplay between the continuous evolution of potentialities and sudden collapse is also an interplay between strict causality and indeterminism. This is a subtle and profound point. Let us look at it in some detail.

Continuous evolution. As long as there is no collapse, the potentiality waves evolve according to a wave equation such as Schrödinger's equation. If we know the shape of the wave at a given time, the equation tells us, precisely, the shape of the wave at any later time. There is no uncertainty, no randomness; the changing shapes of the waves are completely determined from one instant to the next. The subtlety of the issue, however, is this: That which is thus determined does not exist as an actuality in the physical world; it is *the distribution of probabilities for actual events.* Therefore the complete determinism in the evolution of Schrödinger's waves does not eliminate the indeterminism that shows up when we are asked to predict what will actually happen.

As an electron is moving toward a TV screen, we can predict precisely, on the basis of the wave equation, what its probability distribution will be at each instant. We know the chances of it impinging upon the screen at all the different possible locations—we know them *precisely.* Still, what we know precisely is only the chances. We cannot predict where it will hit.

The discontinuity introduced by the collapse. The collapse, the transition from the potential to the actual, is the locus of the indeterminism. A choice is being made as to the place of impingement. This choice is based on the probability distribution: Some locations are more probable than others are, but there is no way to predict the place of impingement.

Let us summarize: That which is continuous, and evolves deterministically, is an underlying field of potentialities. As such it does not exist in the actual world. That which does exist in the actual world, and makes up its elementary constituents, exists as flashes that disappear almost as soon as they appear. These flashes induce discontinuities in the evolution of the potentialities. Moreover, the locations of their appearance are subject to an essential randomness.

It follows that *the actual physical world is made up of units that are discrete not only in space but also in time.* The physical world is made of a myriad of elementary quantum events, flashing in and out of existence, appearing out of a background of potentialities and disappearing back into it, causing, as they do so, discontinuous and unpredictable changes in this background.

2. What Is Matter?

How does this new concept of the structure of physical reality affect our idea of matter? Let us begin by recalling Newton's concept of it:

> It seems probable to me that God in the Beginning formed Matter in solid, massy, hard, impenetrable, movable Particles, of such Sizes and Figures, and with such other Properties, and in such Proportion to space, as most conduced to the end for which he formed them; and as these primitive Particles being Solids, are incomparably harder than any porous Bodies compounded of them; even so very hard as never to wear or break in pieces; no ordinary Power being able to divide what God himself made one in the first Creation.[1]

This concept of matter is the one we still carry intuitively, whether or not we are aware of it. And yet, as we saw in the previous section, the findings of quantum mechanics tell us that matter is something else altogether.

The Newtonian world is a world of objects moving in space. These objects' existence in the actual world is continuous. They affect each other by exerting forces on each other, i.e., by pulling and pushing. The concept of force is central to Newton's physics, and all explanations are derived from it: The force of the horse on the cart is the reason for the cart's motion up the hill. The force of the bat on the baseball is what sends the ball flying through the air.

When we come to quantum mechanics we encounter an entirely different world-view. In particular, *the concept of force is gone.* Furthermore, the actual happenings in the physical world are not explained just in terms of previous actual happenings. They are explained now in terms of a subtle interplay between potentialities and actualities. The Newtonian concept of matter is not only foreign to this world-view, it is downright contradictory to it.

At this stage in our narrative, as we ponder the question of what matter is, we may not be yet in a position to come up with a satisfactory answer. We know, however, that the "substantial" nature of physical reality that we take for granted in everyday life is gone forever. Whatever matter may turn out to be, it is certainly not that.

3. The Loss of Identity, or Identity Is in the Mind of the Beholder

The mystery of the question "What is matter?" is deepened when we consider the question of the identity of atomic and subatomic entities.

Consider two electrons bouncing off each other. They approach each other, and, since they are both negatively charged, they repel each other and scatter away. (You may have noticed that this last sentence seems to contradict the contents of Section 2. In Section 2 we claimed that the quantum world-view does not include the concept of force, and here we are describing two electrons that "repel" each other. The use of the word "repel" is, however, merely a linguistic shortcut. The correct verbal description, the one that emerges out of Schrödinger's equation, does not involve the concept of force, but this description is cumbersome, and irrelevant to the point of the present section.) Let us label the two approaching electrons electron 1 and electron 2, and then look at the two scattered electrons. How can we tell which of the scattered electrons is 1 and which is 2?

The question of identity, or how to tell whether an item you see now is the same item you saw before, is at least as old as Western civilization. It features prominently in the oldest Greek epics known to us, Homer's *Iliad* and *Odyssey*. When Penelope meets Odysseus upon his return to Ithaka, she is faced with an urgent question: How does she tell whether this man who claims to be Odysseus is really her beloved husband? As the story proceeds, this question gets transformed into another: Is there some characteristic of this man, some piece of information, perhaps, that is known only to him and has survived his twenty-year odyssey?

Fortunately for Odysseus and Penelope, and for us, the readers who are eager for a happy ending, such a characteristic is found—the story of how Odysseus built, and where he placed, the marital bed. It is not easy for Penelope to extract this piece of information from him; she does it in a manner that fully justifies the adjective "circumspect," with which the bard often labels her. What a horrible turn would the story have taken if no such decisive piece of evidence were found, and Penelope were tormented by doubt till the day of her death!

As we consider the scattering electrons, we are faced with a similar question: Is there some characteristic of electron 1, or of electron 2, or both, that distinguishes one from the other? And if there isn't, can't we produce one? Can't we put a little dab of red paint on electron 1, or attach a little flag to electron 2?

Alas, the odyssey of the scattered electrons is doomed to a sad ending. The answer to all three questions is a disappointing no. Electrons are very simple entities. They are completely characterized by their mass, electric charge, and spin, and all three quantities are the same for all electrons. The only distinguishing characteristic could be the component of the spin in a given direction, but if the two approaching electrons have the same spin component along some direction, a happy ending is nowhere in sight. Obviously, the idea of adding a marker to an electron is not feasible: Electrons are just about the smallest, simplest particles there are, and even if it were possible to attach something to them, the attachment would not be an innocuous marker.

At this point, a nagging doubt may appear. Are we asking the right question? If there is no way, even in principle, to keep track of the identity of an electron, maybe there is no identity to keep track of; maybe, at the level of elementary particles, the very concept of identity is irrelevant!

Let us discuss this possibility. Suppose the two approaching electrons have just emerged from "electron guns," where their presence was established experimentally (an electron gun is essentially a metal filament that emits electrons when heated). After bouncing off each other they are absorbed by detectors, such as TV screens, where their presence is again established by a measurement of their positions. We are dealing, then, with four elementary quantum events, four "flashes of actual existence" (two at the electron guns and two at the detectors) appearing out of a background of potentialities. According to the formalism of quantum mechanics, when the two electrons are in interaction, their fields of potentialities combine into one. The combined field determines the probability distribution of all the possible configurations of two electrons, but no "record" is being kept as to which electron is which. The very idea of the existence of a continuous identity is just a thought in our minds. Nothing corresponds to it in reality. Identity is in the mind of the beholder.

The following analogy may help: You are sitting in a movie theater, watching an artistic, cartoon-like movie. The producer, a minimalist, experiments with what can be done with two little circles dancing on the screen. You are watching, at first, with only mild interest. After a while, however, you notice something peculiar: When the two circles are far away from each other on the screen, you have no problem keeping them apart. But when they are close to each other, and moving fast, it is no longer possible to keep track. You find this somewhat annoying. They enter into very close quarters and then separate. You wish you could tell which is which!

As you walk out of the theater, the absurdity of the question "Which is which?" hits you. After all, the movie is made of consecutive frames. On each frame the artist drew two little circles. Each circle, drawn in each frame, is new. The question of identity, which annoyed you, is a production of your own mind, born, perhaps, of a wish to see patterns where there are none. The producer, in fact, may have drawn his circles in such a way that you were tricked into looking for such patterns. This does not alter the fact that the movie is made of separate frames and that in reality there is nothing connecting the circles that appear on one frame with the circles that appear on the next. If you watch the same movie in very slow motion, at, say, one frame per second, this becomes obvious.

Strictly speaking, then, in the world of elementary particles there is just one all-inclusive background of potentialities, out of which individual flashes of electrons, protons, photons, etc., make their appearance into momentary actuality. Under certain conditions, e.g., when successive flashes of electrons appear close to each other in a region of space where, at the time, nothing else appears, these successive flashes look like "the same electron." Due to the limitations of language it may even be necessary to use such an expression. Fundamentally, however, the phrase "the same electron" is as meaningless as the phrase "the same little circle" in the movie.

Is this way of looking at quantum processes just a suggestion that one may or may not adopt, depending on one's inclination, or is it for real? After all, it is not easy to accept the idea that electrons really have no identity, that each elementary quantum event is the appearance of a new electron. It turns out, however, that this principle, "Identity is in the mind of the beholder," can be, and has been, proved experimentally! This proof has to do with quantum statistics. The story of the actual experiments is fairly involved, but we can bring the idea across using an idealized, simplified example. While our example is not realistic, it is conceptually equivalent to experiments that have been performed.

Consider two photons moving together in the sense that their probability distributions in space are the same. All photons have spin, and the spin component along the direction of motion can be oriented in one of two directions: It is either in the same direction as the direction of motion or in the opposite direction; that is, the spin component is oriented either "forward" or "backward."

We assume, first, that there is no reason for the spin components to be oriented either forward or backward (i.e., that these two orientations are equally probable), and, second, that the direction of the spin component of one photon has no effect on the direction of the other. For example, if the spin component of one photon happens to be forward, the probabilities of the spin component of the other to be either forward or backward are still fifty-fifty.

Now we ask the following question: What is the probability of the spin components of both photons to be forward?

The answer is not hard to come by. There are four possible configurations (see Table 13.1). Since all four are assumed to be equally probable, the probability of each configuration, including the configuration "both spins forward," is $1/4$.

This is simple and straightforward. *But it is wrong.* Why? It is wrong because the reasoning behind the construction of the table is wrong. We assumed that we are dealing here with two photons, each of which keeps its identity. Because they keep their identity, the configuration "photon 1 forward, photon 2 backward" is different from the configuration "photon 1 backward, photon 2 forward." In reality, however, there is no such thing as "photon 1" and "photon 2." There are only three possible configurations: both forward, one forward and one backward, and both backward. These configurations are equally probable; hence the probability of each is $1/3$.

Table 13.1

Photon 1	Photon 2	Configuration	Probability
forward	forward	forward-forward	1/4
forward	backward	forward-backward	1/4
backward	forward	backward-forward	1/4
backward	backward	backward-backward	1/4

At the beginning of this section we compared the happy ending of the *Odyssey* with the sad ending of the story of the two scattered electrons. Now we realize that the ending of the story of the electrons need not be sad. A happy ending will be regained if we stop looking for what isn't there. The identity of Odysseus is not the identity of the particles that make up his body. It is associated, rather, with the specificity of the complex patterns of interrelationships among these particles, the uniqueness of which can be recognized through the presence of a specific memory, even when, after twenty years, so much has changed. The elementary particles themselves, however, are merely flashes of existence. As such, not only are they devoid of continuous existence, they are not even connected to one another by threads of identity.

Let us summarize what we can say about elementary quantum events, as they are perceived through measurements. These "flashes of actual existence" are atomized in time as well as in space. Furthermore, as they arise out of a background of potentialities and disappear back into it, each one of them has a separate identity. Every elementary quantum event is new.

This is as far as we can go by examining measurements of elementary quantum events. Can we go beyond the knowledge of these events as they are perceived through measurements? Can we know their real nature?

14. The Expression of Knowledge

Can we go beyond the scientific mode of exploration and come to know the real nature of the smallest constituent units of the universe? We learn from Plato that there are different levels of knowledge, which can be accessed through different processes, such as discursive reasoning and contemplation. The conclusions of discursive reasoning, which functions in the subject/object mode, leave room for doubt. Contemplation transcends the subject/object mode; the insights it brings come with "utter certainty." When an insight is formulated, however, the certainty is lost. Nevertheless, the combined results of contemplation and discursive reasoning can lead to the creation of magnificent conceptual structures.

~

The function of insight gives a transcendental content that, when reduced to an interpretive system, becomes subject to the relativity of all subject-object consciousness. Therefore, there can be no such thing as an infallible interpretation. Thus we must distinguish between insight and its formulation.

—Franklin Merrell-Wolff

1. The Experience of the Real, or Knowledge Unlimited

In Chapter 10, Section 5, Julie summarized Heisenberg's statement about his experience in Prunn Castle: "He [Heisenberg] was listening to different speakers suggesting different kinds of order. He felt that 'all these types of order were partial, mere fragments that had split off from the central order,' but, try as he might, he 'was quite unable to find a way to the center through the thicket of conflicting opinions.' At this stage he was engaged in discursive reasoning. And then, something happened. A young violinist started playing Bach's Chaconne, and 'all at once, with utter certainty I had found my link with the center.'"

Heisenberg claims that at that moment he knew something "with utter certainty," but he cannot tell us what he knew. Similarly, James Franck's description of Bohr in a state of contemplation (Chapter 10, Section 5) ends with the statement "Then suddenly one would see that a glow went up in him, and a spark came, and then he said, 'Now I know.'" This phrase, "Now I know," is the end of the story. This ending is extremely significant, because no one among the founding fathers of quantum mechanics liked to talk more than Bohr, and yet, not even he could tell us what it was that he came to know!

What is the reason for this inability to communicate? When we communicate, we do so in the objectivized mode; we communicate about "something." When we come upon a deep truth in the unobjectivized mode of contemplation, and this truth does not lend itself to communication in the subject/object mode, there seems to be no way to tell what we know.

Depth of knowledge and difficulty in communication seem to go together. Heisenberg, one of the most critical and rigorous thinkers of the century, maintained that he found his link with "the central order," and he was utterly certain that he did. Moreover, this claim was not made in the excitement of the experience; it was made thirty years after the event. I see no reason to doubt Heisenberg's implicit claim that sense impressions gathered in the subject/object mode are not the only road to knowledge; *one can be linked to a central order, or come to know the inherent nature of the actual, in a moment of insight, a moment in which the subject/object mode is transcended.*

The phrase "perception in the subject/object mode" means an act of perception in which I am aware of an object. The object can be an external one, such as a star or a flea, or it can be an internal one, such as a thought or a feeling. In either case it is "perception in the subject/object mode" because I, the subject, am aware of it, the object, as an entity that is separate from me.

The issue of the different ways of knowing is addressed by Plato in *The Republic*. Toward the end of the sixth book Socrates introduces distinctions between different types of knowledge. He starts out by explaining to Glaucon the difference between the knowledge of the sensible world and the knowledge of the intelligible realm. He then proceeds to make a further distinction between two kinds of knowledge of the intelligible. The first kind is exemplified by geometry: One starts out with certain hypotheses, or axioms, and proceeds to investigate the consequences. In this mode the axioms themselves are not subject to investigation. As Plato puts it:

> I think you are aware that students of geometry and reckoning and such subjects first postulate the odd and the even and the various figures and three kinds of angles and other things akin to these in each branch of science, regard them as known, and treating them as absolute assumptions, do not deign to render any further account of them to themselves or others, taking it for granted that they are obvious to everybody. They take their start from these, and pursuing the inquiry from this point on consistently, conclude with that for the investigation of which they set out.[1]

Plato proceeds to point out that in such investigations visible forms are used. For example, in order to prove theorems in geometry, one draws triangles, squares, and other geometrical shapes. However, "these things they treat in their turn as only images, but what they really seek is to get sight of those realities which can be seen only by the mind." This first kind of knowing, which we called "discursive reasoning," is then summarized as follows:

This then is the class that I described as intelligible, it is true, but with the reservation first that the soul is compelled to employ assumptions in the investigation of it, not proceeding from a first principle because of its inability to extricate itself and rise above its assumptions, and second that it uses as images or likenesses the very objects that are themselves copied and adumbrated by the class below them [i.e., visible forms].

This first kind, or "class," as Plato calls it, is then contrasted with the second and higher kind of knowing, which we called "contemplation":

Understand then, said I, that by the other section of the intelligible I mean that which the reason itself lays hold of by the power of dialectic, treating its assumptions not as absolute beginnings but literally as hypotheses, underpinnings, footings and springboards, so to speak, *to enable it to rise to that which requires no assumption and is the starting point of all,* and after attaining to that again taking hold of the first dependencies from it, so to proceed downwards to the conclusion, making no use whatever of any object of sense but only of pure ideas, moving on through ideas to ideas and ending with ideas [italics added].

When one is engaged in discursive reasoning, one functions in the subject/object mode. The visible forms that are used as images, and the intelligible ideas about them, are objects in front of one's mental eye. For example, while thinking about Pythagoras's theorem, one visualizes a right-angled triangle and thinks about the lengths of its sides. When one engages in contemplation, however, one functions differently. Rather than looking at a mental object, one rises "to that which requires no assumption and is the starting point of all." Transcending the subject/object mode, one finds one's link with the central order, as Heisenberg puts it.

Pondering this higher mode of knowing, we may well respond to Socrates's explanation as Glaucon did: "I understand, he said, not fully, for it is no slight task that you have in mind."

If it is indeed possible to come to know deep truths through such a higher mode, is it really impossible to communicate them? Well, it depends on what we mean by "communication." All day long we communicate information, thoughts, and feelings. The communication of deep truths, however, cannot take place in the same way. The reception of such truths requires that the receiver is able to make contact with them, that he or she is open to the transcendence of the subject/object mode. Such communication, then, is not communication of information; it is communication of an experience. Even if such communication is possible in principle, however, it may well be impossible in practice because of the limitations of the receiver. This last point is neither profound nor hard to accept:

After all, if Heisenberg wished to communicate to someone not what he came to know in Prunn Castle but, rather, the mathematical formalism of quantum mechanics, would he be able to do so? Years of training in mathematics are required to understand what he would say, and most of us are unprepared.

Great discoveries in science and great works of art are grounded in transcendent experiences. In some sense they point the way back to the experiences that are at their source. Consider, for example, Mozart's experience of writing a symphony. His experience must be very different from my experience when I listen; yet something of his experience is conveyed to me, especially if, as I listen, *my best attention is present*. Even then, however, there is a profound difference, a difference in levels, between Mozart's experience of writing a symphony and my experience of listening to it. Accounts of great artistic creations, as well as scientific discoveries, indicate that moments of such creations and discoveries are moments of contact with the noumenal realm. My experience of listening occurs, more or less, in the subject/object mode: As I listen, most of the time the music is felt to be "out there," while I am receiving it "here." I do, however, need to add the qualifiers "more or less" and "most of the time:" When my attention is both very relaxed and very concentrated, there are moments when the subject/object mode disappears. These moments bring joy. It is as if they represent a thread, however thin, connecting me to the experience of the music behind the music—the real music behind the notes that are heard.

It seems, then, that the score of a Mozart symphony represents *a projection of an experience from the level of unobjectivized unity onto the level of the ordinary, space-and-time-bound experience—experience in the subject/object mode*. It is a kind of a "spilling over," from the level of unity to the level of the experience of phenomena. It conveys something of the unity-experience, or the unity-knowledge, but only as much as the lower level can contain. From its side, this lower-level experience can indicate the direction toward the higher one, and, when one is prepared, and one's attention is acute enough, some thread connecting the two kinds of experience can be established.

2. Formulation of an Insight, or Knowledge Limited

The principle we have just discussed in relation to music seems to apply just as well to experiences of insight in science. An experience of insight transcends the subject/object mode; one cannot describe it, and yet it can give rise to a scientific theory, one that is closely related to the experience. This is exemplified by Heisenberg's account of his discovery of the uncertainty principle.

Let us recall the crucial steps of the process. During his stay at Bohr's institute in Copenhagen, Heisenberg struggled to resolve an apparent contradiction between the mathematical formulation of quantum mechanics and an experimental fact that seemed to be at odds with it. The mathematical formulation did not allow for the existence of continuous trajectories of electrons, yet, when he looked at photographs of a cloud chamber, "the path of the electron through the cloud chamber obviously existed." This struggle went on for months. And then:

> It must have been one evening after midnight when I suddenly remembered my conversation with Einstein and particularly his statement, "It is the theory which decides what we can observe." *I was immediately convinced that the key to the gate that had been closed for so long must be sought right here.* I decided to go on a nocturnal walk through Faelled Park and to think further about the matter. We had always said so glibly that the path of the electron in the cloud chamber could be observed. But perhaps what we really observed was something much less. . . . In fact, all we do see in the cloud chamber are individual water droplets which must certainly be much larger than the electron. The right question should therefore be: Can quantum mechanics represent the fact that an electron finds itself approximately with a given place and that it moves approximately with a given velocity, and can we make these approximations so close that they do not cause experimental difficulties?
>
> A brief calculation after my return to the Institute showed that one could indeed represent such situations mathematically, and that the approximations are governed by what would later be called the uncertainty principle of quantum mechanics [italics added].[2]

According to Heisenberg's account, *the breakthrough and the formulation of the discovery are different events, well separated in time.* The breakthrough came when Heisenberg suddenly remembered Einstein's statement "It is the theory which decides what we can observe" and is expressed by the sentence "I was immediately convinced that the key to the gate that had been closed for so long must be sought right here." Following the experience, Heisenberg engaged in further thought, as well as some calculations, and only then was he able to formulate the uncertainly principle. This is a beautiful illustration of the relationship between the unobjectivized experience of insight and its projection onto the paradigm of the culture. Heisenberg's account also illustrates the fact that to come to a *formulation* of a deep insight, the *experience* of insight is not enough—the formulation is not created in a vacuum. In the case of science some element, or elements, of the ineffable insight must be expressed in words or formulas, and, furthermore, the formulation must agree with known facts and be coherently related to the formulations of other relevant insights. In Heisenberg's case he had to verify that his newly formulated principle was not in contradiction with either the cloud chamber pictures or whatever thought experiments he could conjure up.

At the end of the last chapter we raised the question "What is the real nature of elementary quantum events? What are they in themselves, independent of the perceiver?" If the totality of our experiences were limited to perceptions and measurements in the subject/object mode, the answer would have been "There is no way to know." The results of the present chapter lead us, however, to a different conclusion.

An experience of insight can lead to a direct knowledge of reality as it is in itself, a knowledge that comes with the feeling of utter certainty. This knowledge is not expressible in conceptual terms. A conceptual formulation communicates an aspect, or aspects, of it. Even if such a formulation ushers in a paradigm shift, the expression of the shift is bound up with language, culture, and the many paradigms implicit in them. The formulation does not share the utter certainty of the experience.

3. The Place of Speculative Thought

Is it useless, then, to engage in ontological discussions, i.e., to try to express conceptually what something really is, from its own side? Is it a waste of time to try to answer the question "What is the real nature of elementary quantum events?" Not in the least. Just as an attentive listening to a Mozart symphony can establish a thread with the noumenal experience that is its source, a thoughtful absorption in trying to understand the formulation of an insight can bring us closer to the experience of the insight and to the utter certainty that came with it.

There are many accounts of transcendent, insightful experiences that led to important discoveries. The formulations of discoveries, however, involve a weaving of the expressions of such insights together with other relevant insights, theories, and facts into coherent conceptual structures. This process involves speculation. Hypotheses that emerge as a result of such a process often become unifying elements in the creation of a whole system of thought. This approach has been employed by Alfred North Whitehead in the formulation of his celebrated "process philosophy," which will be explored in the next chapter.

The creative combination of insights, speculations, and discursive thinking may, but need not, lead to ontological descriptions. For Heisenberg it led to an ontological interpretation of quantum mechanics. For Niels Bohr, however, it led to a denial of the existence of a "quantum world" and to the introduction of complementary modes of descriptions of the "whole phenomena" that arise in quantum measurements. This shows that the certainty of the experience of insight does not lead to certainty in the conceptual formulations that result from it.

Whitehead's "process philosophy" is not an interpretation of the quantum theory. It is, rather, a comprehensive philosophical system that turned out to be in surprising agreement with the findings of quantum mechanics. Its doctrines will help us understand more deeply the nature of elementary quantum phenomena and their role as "atoms of reality." Before we turn our attention to Whitehead, however, let us visit with our old friends, Peter and Julie.

4. Peter's Wrath

My next meeting with Julie and Peter took place in Paris, where they were attending an International Symposium on Statistical Methods for the Evaluation of

Probabilities of Space Debris Impact on Spaceships in Interstellar Voyage. Our meeting was brief; I simply said hello and gave them copies of the draft of the new parts of this book, the chapters I had written since our last meeting. We decided to meet at a sidewalk café the next day.

As I approached the café Julie and Peter were already there. Peter's back was turned toward me, but I could tell from his gestures that he was in quite a state, and Julie was trying to calm him down. As I came closer, I could make out what he was saying.

"Absolutely not! This is totally unacceptable! I mean, who does he think he is? Don't you see how stupid and arrogant this is? I'll have none of it!"

"Oh, Peter!" Julie had just seen me. She did not want Peter to hurt my feelings. I came to the table and greeted them as best I could, wishing to find out what the fuss was all about. I did not have to wait long.

"Oh, there you are!" Peter was not even trying to hide his contempt. "Well, what is this?" He was holding in his hand the pages I had given him the previous day. "This is garbage! I can't believe you have actually written this! I'm getting out of here. I will no longer be a part of this book of yours!"

I was both hurt and amused, and I could not tell which reaction was stronger. "My dear Peter," I said, "before you take such a drastic step, will you calm down enough to tell me what's making you so mad?"

"All right. Sure, I'll tell you." He calmed down just a bit. "If you publish this, you will be the laughingstock of the world of science. Look at this last chapter, Chapter 14. In one short section, Section 1, you claim to solve the central problem of epistemology! In another short section, Section 2, you similarly 'solve' the issue of transcending the subject/object mode! And in Chapter 12 you try to convince me that this table is alive! My friend, you are a good physicist. You understand the principles of quantum mechanics very well. As long as your writing was confined to that, you did okay. Why don't you keep to your area of competence?"

"You mean physics?"

"Not even that. I mean your area of competence within physics. For example, you probably couldn't even give a sensible lecture in this symposium Julie and I are attending, because you know nothing about statistical methods for the evaluation of probabilities of space debris impact on spaceships in interstellar voyage."

"True," I said, "but this is only because I never took the time to study the subject. Had I been interested in it, I would have applied my knowledge of physical principles and mathematical methods to it and would have presented your symposium with a damn good paper."

"Sorry. You're right about that." One could always count on Peter's honesty, even when he was mad. "But my point is this: Throughout the history of science, from the ancient Greeks all the way to the Renaissance, people seem to have believed that every scientist can know everything there is to know. Often a scientist was a physician, a theologian, an astrologer, a mathematician, a physicist, and much else besides. Sometime during the nineteenth century people came to the conclusion that the amount of scientific knowledge is so vast that scientists need to specialize. And it worked. Specialization is one of the secrets of scientific progress in the past two centuries. You seem to have taken no notice of that. Your book includes sections on philosophy, poetry, psychology, even music! I repeat my question: Why don't you stick to your area of competence?"

The atmosphere around our little table was rather tense, and I must admit that I felt at a loss. Peter's point was well taken. Why was I writing as I did? Was I walking, unawares, into an intellectual minefield? I was grateful to the waiter who came to take our orders just at that moment—he gave me a little more time to think. But Julie, who seemed to understand my motivation, spoke up.

"Look, Peter," she said, "you've just extolled the virtues of specialization, and you were quite right to do so. But haven't we learned something about complementarity? Don't you see that specialization by itself leads to catastrophe? Economists make the economy stronger, but, being specialists, they are unaware of the destruction of the environment that their actions often cause. Environmentalists are appalled by what's happening, but, not being psychologists, they don't know how to bring their message to the public without arousing people's anger. Social workers see the misery of poor people, but, not being experts in systems theory, they often provide solutions that are worse than the problems. And so on. You see what I mean."

"Yes, I guess I do," Peter admitted. "But what do we do now? What can replace specialization?"

"Well, I just mentioned complementarity. Specialization is here to stay, but it needs to be balanced by a holistic approach. Otherwise specialists, who are unaware of the limits of validity of their abstractions, will continue mistaking them for concrete facts. As you can ascertain for yourself, very few people are even aware of the distinction between abstraction and concrete fact. Paradigms are adhered to as if they were absolute truths."

"In the case of physics the situation has become really ludicrous," I added, having regained my balance. "The Newtonian paradigm was problematic philosophically right from the start. For example, as I explained in Section 6 of Chapter 12 , in a clockwork universe it makes no sense to hold people responsible for their actions. So one would have thought that people would have been happy to be rid of it. But no! At this point in the development of physics we know that the Newtonian conceptual framework is all wrong. Relativity and quantum mechanics have been expounded in hundreds of books. But, somehow, the paradigm of the culture is still largely Newtonian."

"So what are you trying to do? Add another book to the list?" Peter's attitude towards me still hadn't really changed.

"No. I'm trying to get three messages across. First, that the paradigm based on local realism is defunct even within physics. And since physics was its only raison d'être, it does not make any sense to hold on to it. This in itself entails a revolution. For example, the reductionistic method of explanation can no longer be considered the only valid one. Second, that while physics does not provide a complete new paradigm, it does indicate a direction. Indeterminism, the framework of complementarity, the holistic aspect of systems, the existence of which was proved by John Bell, the rise of the concept of 'potentiality' to a new level of significance, and the understanding of the elements of physical reality as 'flashes of existence'—all of these are powerful indications, showing the direction toward the new, emergent world-view. And third, I wish to supplement the findings of physics by indications from other realms of human experience: philosophy, poetry, music, records of intuitive insights. These seem to give new insights, insights that are not provided by physics but certainly do not contradict it in any way. The theme of Chapter 12, the aliveness of nature, is one example of such an indication. Schrödinger's exposition of the principle of objectivation is another. All these elements, taken together, can help us in the search for a new paradigm. And a search for a new paradigm—make no mistake about it—is a search for a new way of life."

"I can appreciate that." Julie joined the conversation, addressing both of us. "Even the little bit I've learned so far has affected my thinking, as well as my feelings. I was taken, for example, by the idea of complementarity. This radically changed my attitude to expressions of deep truths. I used to take every such expression as an absolute statement. Now I tend to take such an expression as relative— as one of two complementary expressions. Bohr's statement to the effect that often the opposite of a deep truth is another deep truth made a strong impression on me."

"What do you mean by that? Can you give me an example?" Peter was really interested now. The expression on his face lost its hard lines.

"Well, yes. As a teenager I was impressed by the statement 'Every man is an island.' It sounded so true, and it was such a pithy expression of the loneliness I felt. A little later I came across the statement 'No man is an island.' It too felt true, and I became confused, not knowing which of these two statements was really true. This apparent contradiction was vaguely present at the back of my mind all these years. The idea of complementarity provided me with a resolution: Both statements are true. They apply in mutually exclusive conditions; hence there is no contradiction between them."

"What do you mean, 'they apply in mutually exclusive conditions'?"

"We astronauts know very well what it is like to feel totally alone, totally 'an island.' Remember when your spaceship, coasting between Mars and Jupiter, was hit by a sizable piece of debris, and ten different problems needed your immediate attention, all at once? At that moment you exemplified the statement 'Every man is an island.' But then there are the other moments, moments that are very different." Julie's voice became softer. "I remember, for example, a moment of looking into your eyes, knowing, with certainty, that neither of us is an island."

"Yes, I remember the moment." Peter was quite mellow. I was impressed by the change that had come over him in the short span of time since I entered the café. Maybe Empedocles was right, I thought, when he considered Love as a cosmological principle. I enjoyed listening to Peter as he picked up the thread.

"Come to think of it," he said, "I too went through some changes while thinking about a paradigm shift. Following our conversation about the principle of objectivation, I became really interested in finding out what it would be like *not* to exclude the Subject of Cognizance from the field of my attention. 'What would it be like,' I asked myself, 'to have an attention without an object, to let the attention delve into the unobjectivized self?' I'm still at the initial stages of this exploration. I'm finding out, for example, that it is almost impossible to come to the point of having such an attention, but I've glimpsed an intimation that, with a proper investment of time, patience, and intelligence it will be possible. At any rate, Schrödinger's suggestion has certainly had a profound effect on me."

We sat quietly. Each was absorbed in his or her own thoughts, yet we were somehow connected. I was not surprised when, after a few minutes, I heard Peter's voice addressing me.

"As you can probably tell," he said, "I'm not angry anymore. But I am still bothered. Do you really believe that you resolved the central issues of epistemology and ontology in the first two sections of this chapter?"

"I don't know," I responded softly. "I do believe that what I have written in these sections is true. And I am sure I was not the first to express these ideas. There may or may not be some novelty in the formulation of these thoughts, but their essence has been known for a long time." I paused, then continued, looking him in the eye. "There is something else we have to settle. Are you still mad at me for implying that the table is alive?"

Peter seemed amused, and then a slight frown appeared on his face. "I'm not mad," he said, "but do you really believe it?"

"I believe that the universe and all its constituents are alive," I responded, "but there are degrees of 'aliveness.' Take your body as an example. Your hair is alive, and so is your heart. But your heart is much more alive than your hair. It is impossible to consider your heart an inanimate object, but it takes training to discern the life in your hair. As a matter of principle, though, your whole body, and every part of it, is alive. I see the question of this table in a similar light. This cheap, metal and plastic, mass-produced table is not very alive, just like your hair. Other tables, produced by master craftsmen, may be different. But such questions of gradation, and the application of this gradation to individual objects, have little bearing on the principle."

"Well," Peter said, "the paradigm of the living universe is certainly appealing. It warms your heart. I wish I could accept it."

It was Julie who had the last word. "And what's in the way of your accepting it, Peter?" she asked. "Look at it. It is well worth finding out."

15. A Universe of Experience

According to Whitehead's "process philosophy" we live not in a universe of objects but in a universe of experiences: The basic constituent units of the universe are "throbs of experience." How does this idea compare with quantum mechanics? A complete agreement between Whitehead's philosophy and quantum mechanics cannot be expected, because the latter is subject to the principle of objectivation, which excludes experiences from its domain of inquiry. It turns out, however, that elementary quantum events come as close to Whitehead's "throbs of experience" as objectivized entities can.

~

[The Intellect:] Ostensibly there is color, ostensibly sweetness, ostensibly bitterness, actually only atoms and the void.
[The Senses:] Poor intellect, do you hope to defeat us while from us you borrow your evidence? Your victory is your defeat.

—Democritus of Abdera

1. Occasions of Experience

What is the real nature of elementary quantum events? What are they in and for themselves? We are finally in a position to suggest an answer to these questions, an answer based on Alfred North Whitehead's "process philosophy."

If the universe is alive, does it not stand to reason that its basic units, the units that show up in the context of quantum mechanics as elementary quantum events, are the basic acts of life, elementary experiences? According to Whitehead, the ultimate building blocks of reality are discrete items, which he calls, synonymously, "occasions of experience," "throbs of experience," "actual occasions," or "actual entities," depending on the context. He asserts: "I hold that these unities of existence, these occasions of experience, are the really real things, which in their collective unity compose the evolving universe, ever plunging into the creative advance."[1]

The paradigm shift suggested by Whitehead is radical: The ultimate "atoms of reality" are experiences! This is very different from the current world-view, which is based on the principle of objectivation. According to the latter, the basic constituents of reality, the so-called elementary particles, are inanimate bits of

163

matter. Large units of matter that are made up of these particles, namely, stars and planets, are also inanimate. Living things are insignificant on a cosmological scale; hence experiences, which only living things may have, are equally insignificant. What could have motivated Whitehead to discard the current paradigm so completely and introduce the notion that experience is at the very center of everything that exists?

To the best of my knowledge, Whitehead never discussed how he discovered and created his system. He presents the system as a coherent, logical one, formulated "to obtain a self-consistent understanding of things observed."[2] His pupil Victor Lowe, however, discusses Whitehead's placement of experiences at the center of his system at some length. Lowe believes that the source of this placement is both intuitive and analytical.

The intuitive motive is this: "We instinctively feel that we live in a world of 'throbbing actualities'; and such 'direct persuasions' are the ultimate touchstone of philosophic theory."[3] As we saw in Chapter 12, such intuitions are shared among the more sensitive poets, artists, and scientists.

Analytically, the assumption that experiences are the primary elements of the universe leads to a resolution of the dichotomy that has plagued Western philosophy since Descartes: the dichotomy of mind and matter. According to Descartes the elements of actual existence are of two kinds: material and mental. These types of existence are different and incommensurate: The table that I see in front of me is material, while my intention to go on typing is mental; the two have nothing in common. This duality of mind and matter creates enormous difficulties. For instance, how can these utterly different items be related? How does my intention to lift my arm (a mental event) bring about the actual lifting of the arm (a material event)? Whitehead's brilliant analysis of these difficulties led him to the conclusion that a self-consistent paradigm must be based on the hypothesis that there is only one basic kind of actual existence, and if there is only one kind, it must be of the nature of experience: The fact that experiences exist cannot be denied. Not only are we certain that we do experience, but everything we believe we know about the universe, including the materialists' belief in the primacy of matter, is deduced from our experiences. As the epigraph of this chapter demonstrates, this truth was understood by Democritus twenty-five centuries ago! Lowe explains:

> The assumption of the primacy of experience resolves not only the dualism of mind and matter; it resolves the dualism of fact and value as well. Whitehead does not wish to think that intrinsic value is an exclusive property of superior beings; rather, it belongs to even "the most trivial puff of existence." In human life, he finds value not far off, but at hand as the living essence of present experience. If every puff of existence is a pulse of some kind of immediate experience, there can be no final dualism of value and fact in the universe.[4]

Finally, Lowe cites Whitehead's "love of concrete immediacy" as another motivating factor.[5] Whitehead was keenly aware of the distinction between abstractions and concrete facts. It was natural for him to discard abstractions, such as the independent existence of material objects, and choose the only concrete facts we know, namely, experiences, as the basis of his system of thought.

2. What Is an Experience?

Let us probe a little deeper into the Whiteheadian world-view. If throbs of experience are the basic elements of reality, what, in Whitehead's terminology, is "an experience"?

We will begin addressing this question by clarifying what an experience is not. An experience may, but need not, be *conscious*. For example, when Julie was talking to Peter about his driving, she pointed out that often he would be driving for a long time without being conscious of himself, his surroundings, or his handling of the wheel. While Peter was driving, he was experiencing all the time. He was perceiving things, gauging distances, making decisions; yet he was not conscious of his experiences. This distinction between conscious and unconscious experience is crucial. When Whitehead applies the concept of "experience" to animals, plants, and putatively inanimate things, he does not mean that they are conscious of their experiences; he only means that they do have them.

But what does it mean to have an experience? To begin, the very expression "to have an experience" is too abstract. The separation between "myself," or "the one who has the experience," and "the experience itself" is a mental construct that does not correspond to the fact. The experience itself is the concrete fact—*being* an experience rather than having one. Whitehead's "throbs of experience" are just that— *units of experience that create their subject as they create themselves.*

At long last we are ready to answer the question "What is an experience?" A "throb of experience" is a process that involves, first, "individual self-enjoyment of a process of appropriation," second, the "immediacy of self-enjoyment" in "the transformation of the potential into the actual," and, third, the presence of "aim" in this process of self-creation.[6] Let us elaborate.

An experience arises in the context of an already existing universe. It must "appropriate," or relate to, whatever has taken place already. For example, as I listen to a piece of music, my experience at any moment "accommodates" or "appropriates" the sounds produced a split second before. The state of my body, musical education I may or may not have had, the acoustics of the room, etc., etc.—all are factors from the immediate and distant past that contribute to the make-up of the experience. *How* they enter into the experience depends on how I appropriate them.

At any moment, in addition to all these appropriations (also called "prehensions"), an experience involves "the transformation of the potential into the actual." In the example of listening to music, this transformation depends on my attention. While the music goes on, I have the potential to listen to it, as well as the potential to engage in any number of other activities. If I do listen, I can try to listen for a particular instrument, enjoy the melodic line, compare the piece I hear with other pieces I remember, and so on. The range of potentialities is unlimited, but very few of these potentialities become actual in the experience.

The choice of what potentiality to actualize involves the third aspect, namely, "aim." The data that have been chosen for actualization are not being actualized as a mere collection of diverse items; rather, they are organized into a structured whole. For example, if I listen to a symphony, I "appropriate" the sounds of the various instruments that are playing, but I don't appropriate them "side by side." Rather, I appropriate them in an interrelationship that creates in me the experience of listening to that symphony. In doing so, I am following the "aim" of the experience. "The aim," says Whitehead, "is at that complex of feeling which is the enjoyment of those data in that way."[7]

This brings us to the center of our description of these "throbs of experience": While each one of them appropriates the many diverse pieces of data contained in "the antecedent functioning of the universe," *each "throb of experience" is one. Its essential feature is the self-enjoyment of the unification of the many into one.* This "one" is new—a new way of unifying the many. Hence the drama of the activity of the universe is the drama of "creative advance into novelty."[8]

The example I have chosen, listening to music, involves a series of conscious experiences. Whitehead's characterization of the nature of experience, however, applies to the unconscious as well as the conscious domain. It applies, for example, to Peter's experiences during his trip while he was driving unawares. And Whitehead applies the same characterization not only to the experiences of animals, plants, and bacteria but also to those of any other entity—specks of dust, stars, the universe as a whole. All the entities of nature are what he calls "societies" of "throbs of experience."

Let us conclude this introductory exposition of Whitehead's "actual entities," or "throbs of experience," by commenting on his repeated use of the expression "self-enjoyment." First of all, Whitehead does not mean that actual entities are necessarily conscious of their "self-enjoyment." (While Peter was driving unawares, he may well have enjoyed looking at an antique car he happened to pass, without being conscious of his enjoyment.) Second, the term "self-enjoyment" is used in a generalized sense; it indicates that *each actual entity has an absolute value in itself and for itself.* Third, just like the word "experience," the term "self-enjoyment" points beyond its literal meaning and indicates a direction of reasoning and contemplation. This point is the subject of the next section.

3. The Reorganization of Thought

All verbal definitions and explanations are relative: They define or explain one concept in terms of other concepts. In employing them we create a new meaning by invoking a combination of existing meanings. If a new idea is revolutionary enough, however, its reception by the reader calls for more than a new set of connections among old meanings. It calls for a structural change in the patterns of his or her thinking.

Fortunately, we do have the innate capacity to modify the patterns of our thinking in response to encounters with new experiences. Such modifications are the results of the confrontation between the old paradigm and the new material. Feelings of uneasiness are an essential part of the process and have been appropriately called "creative frustration." As the psychologist Jean Piaget discovered while studying the intellectual development of children, the movement toward *assimilation* of the new material into the old paradigm is complemented by another movement, one that tries to *accommodate* one's mode of thinking to the new material, and the result of these two complementary movements is a new inner *organization*—the appearance of a new paradigm of thought.

The intellectual development of children involves transitions from one phase to another. Piaget demonstrated that each of these transitions is marked by a reorganization of the child's mental capacities, whereby new ways of thinking become possible. Such reorganization cannot be taught directly, since anything that one tells the child is assimilated into his or her existing modes of thought. The required reorganization arises instead out of the child's activity, as he or she tries to *assimilate* new experiences into his or her current mental framework, while simultaneously trying to *accommodate* the framework to the new experiences.

If we wish to understand the Whiteheadian ideas of "experience" and "self-enjoyment," we must engage in a similar process. Whitehead's explanations involve the use of ordinary words, such as "appropriation," "actual," and "occasion," but these words are used in a generalized sense, thereby indicating a new way of thinking. Such indications are all that one can provide. Once they are given, it is up to us to try, through both discursive reasoning and silent contemplation, to allow the suggested reorganization to occur in us.

In my own experience, whenever I think about the terms introduced by Whitehead, old modes of thought keep reasserting themselves. For example, when I think about "an actual entity," I can't help having a blurred image of a "something" in front of my mind's eye. I have to keep remembering that an actual entity is not a "thing," that it is, as we will see in the next section, an atemporal process of self-creation. To be sure, a collection of actual entities, such as the one I call "my computer," may *appear* to me as a "thing." This appearance, however, is not what the collection is in itself; it is, rather, the result of my objectivation of this collection of actual entities.

4. The Cosmological Function of Feelings

"'Actual entities,'" Whitehead writes, ". . . are the final real things of which the world is made up. There is no going behind actual entities to find anything more real. They differ among themselves: God is an actual entity, and so is the most trivial puff of existence in far-off empty space. But, though there are gradations of importance, and diversities of function, yet in the principles which actuality exemplifies they are all on the same level."[9]

When we discussed the process of the collapse of a quantum state, we emphasized its atemporal nature. We saw that the process of collapse does not take time; rather, its *completion* is marked by the appearance of an elementary quantum event in spacetime. Such an event is brief: It is over almost as soon as it happens.

An actual entity's relation to time is similar to that of a collapse: An actual entity is an atemporal process of self-creation. Its "genetic" growth from phase to phase is a complex process that does not take place in time. When the process is complete, the completed entity appears in spacetime—and perishes almost as soon as it appears. By perishing, however, it becomes immortal: The fact that it did occur can never be eradicated. *It has taken place*, and hence it is "a stubborn fact," which every subsequent actual entity has to appropriate.

Each actual entity is a discrete item, separate from all the other actual entities. This statement seems to contradict the obvious: Mountains endure for millions of years; flowers endure for weeks; each one of us has a sense of being the same person over decades. If each actual entity is discrete and almost instantaneous, how do we account for this apparent continuity in nature and in ourselves?

Whitehead regards mountains, flowers, and people as "societies," or successive collections, of actual entities. At each new moment the mountain is a new collection. It seems to be a (relatively) permanent entity because the successive collections of actual entities that embody it are very much alike. Similarly, when I listen to a piece of music, the same note may endure for a few seconds. In these few seconds I have a number of experiences that are very much alike: The environment out of which they arise stays the same, and each experience, except the first one, "imitates" the one preceding it. And since my attention is too coarse to detect the granular nature of this chain of actual entities, they seem to me to constitute a continuous whole—just as, while flying over sand dunes in an airplane, I see smooth contours and cannot discern the granular nature of the sand, and just as, while watching a movie, I see the characters moving continuously and I don't discern the discrete nature of the sequence of frames.

Let us turn to a different issue: We have repeatedly referred to one actual entity "appropriating" or "prehending" other actual entities. How does it do that? What is the nature of the relationship between a given actual entity, which is the process of its own self-creation, and the actual entities it prehends in the process?

To answer this question we need to make a detour into an exploration of perception. Appropriation or prehension is a generalized analogue of human perception; and, conversely, human perception is an instance of prehension, the only instance we are intimately familiar with. Whitehead's general characterization of prehensions is based on an insightful analysis of human perception.

Whitehead realized that perceptions are based on "an indefinite set of obscure bodily feelings that form a background of feeling with items occasionally flashing into prominence."[10] These "obscure bodily feelings" are the *means* whereby we perceive the entities presented to us by the senses. Consider the sense of touch: When I touch the tabletop and perceive its texture, what I really perceive are "obscure bodily feelings" of my fingertips. These are conveyed to the brain, which objectivizes them and constructs a percept, the texture of the tabletop. The same principle applies to the other senses as well. For example, we see with the eyes. This means that the obscure, unconscious experiences of "feeling the eyes" are behind the (conscious or unconscious) experience of seeing something.

Additionally, the contents of our perceptions are permeated through and through with emotive influences. When we look around, we see a display of shapes and colors, but colors and shapes affect us emotionally: Red arouses, green soothes, blue induces cool feelings; circles affect us harmoniously, sharp edges disturb us. When we listen, we are comforted by harmonious melodies and annoyed by screeching sounds. These emotions are induced both directly, i.e., physiologically, and indirectly, through the memories of analogous experiences in the past. In short, our perceptions are based on and bathed in an ocean of subtle feelings and emotions of which we are ordinarily only vaguely aware.

Generalizing this observation, Whitehead asserts that an actual entity "prehends" the other actual entities that exist as "stubborn facts" in its universe by "feeling" them. "There is nothing in the real world," he says, "which is merely an inert fact. Every reality is there for feeling: it promotes feeling, and it is felt."[11]

Victor Lowe is right, I believe, when he warns us of "the danger of reading too much into the term 'feeling' . . . to be 'felt' means to be included in an integrative, partly self-creative atom of process, as part of the internal essence of that process."[12] Feeling is a relationship between the one who feels and that which is felt. It is this meaning of the word "feeling" that is generalized to encompass the relationship of any actual entity as a process of self-creation with any other actual entity that exists in its universe as "a stubborn fact."

5. "The Elucidation of Things Observed"

At this point in our narrative,

> the question at once arises as to whether this factor of life in nature, thus interpreted, corresponds to anything that we observe in nature. All philosophy is an endeavor to obtain a self-consistent understanding of things observed. Thus its development is guided in two ways, one is demand for coherent self-consistency, and the other is the elucidation of things observed. It is therefore our first task to compare the above doctrine of life in nature with our direct observations.[13]

How are we to conduct such a comparison? Should we turn to science? No. There is no way in which the scientific endeavor can detect the aliveness of things: Its methodology rules out the possibility of such a finding. Let us quote Whitehead on this point:

> Science can find no individual enjoyment in nature; science can find no aim in nature; science can find no creativity in nature; it finds mere rules of succession. These negations are true of natural science. They are inherent in its methodology. The reason for this blindness of physical science lies in the fact that such science only deals with half the evidence provided by human experience. It divides the seamless coat—or, to change the metaphor into a happier form, it examines the coat, which is superficial, and neglects the body which is fundamental.[14]

Whitehead claims that the methodology of science makes it blind to a fundamental aspect of reality, namely, the primacy of experience. It neglects half of the evidence. Working within Descartes's dualistic framework of matter and mind as separate and incommensurate, science limits itself to the study of objectivized phenomena, neglecting the subject and the mental events that are his or her experiences.

Both the adoption of the Cartesian paradigm and the neglect of mental events are reason enough to suspect "blindness," but there is no need to rely on suspicions. This blindness is clearly evident: Scientific discoveries, impressive as they are, are fundamentally superficial. Science can express regularities observed in nature, but it cannot explain the reasons for their occurrence. Consider, for example, Newton's law of gravity. It shows that such apparently disparate phenomena as the falling of an apple and the revolution of the earth around the sun are aspects of the same regularity: gravity. According to this law the gravitational attraction between two objects decreases in proportion to the square of the distance between them. Why is that so? Newton could not provide an answer. Simpler still, why does space have three dimensions? Why is time one-dimensional? Whitehead notes, "None of these laws of nature gives the slightest evidence of necessity. They are [merely] the modes of procedure which within the scale of observations do in fact prevail."[15]

This analysis reveals that the capacity of science to fathom the depth of reality is limited. For example, if reality is, in fact, made up of discrete units, and these units have the fundamental character of being "throbs of experience," then science may be in a position to discover the discreteness; but it has no access to the subjective side of nature, since, as Schrödinger points out, we "exclude the Subject of Cognizance from the domain of nature that we endeavor to understand." It follows that in order to find "the elucidation of things observed" in relation to the experiential or aliveness aspect, we cannot rely on science; we need to look elsewhere.

If, instead of relying on science, we rely on our immediate observation of nature and of ourselves, we find first that this [i.e., Descartes's] sharp division between mentality and nature has no ground in our fundamental observation. We find ourselves living within nature. Secondly, I conclude that we should conceive mental operations as among the factors which make up the constitution of nature. Thirdly, that we should reject the notion of idle wheels in the process of nature. Every factor which emerges makes a difference, and that difference can only be expressed in terms of the individual character of that factor.[16]

Whitehead proceeds to analyze our experiences in general, and our observations of nature in particular, and ends up with "mutual immanence" as a central theme. This mutual immanence is obvious in the case of human experience: I am a part of the universe, and, since I experience the universe, the experienced universe is a part of me. Whitehead gives an example: "I am in the room, and the room is an item in my present experience. But my present experience is what I am now."[17] A generalization of this relationship to the case of any actual occasion yields the conclusion that "the world is included within the occasion in one sense, and the occasion is included in the world in another sense."[18] The idea that each actual occasion appropriates its universe follows naturally from such considerations.

The description of an actual entity as being a distinct unit is, therefore, only one part of the story. The other, complementary part is this: The very nature of each and every actual entity is one of interdependence with all the other actual entities in the universe. Each and every actual entity is a process of prehending or appropriating all the other actual entities and creating one new entity out of them all, namely, itself.

6. The Evidence of Quantum Mechanics

The factor of experience, or aliveness, cannot be revealed by scientific methods. There are, however, other aspects of Whitehead's system that can be subjected to scientific scrutiny. In the present section we will discuss a remarkable correspondence between fundamental Whiteheadian ideas and the discoveries of quantum physics. This correspondence is especially remarkable if we bear in mind that when Whitehead formulated his philosophical system he was unaware of Heisenberg and Schrödinger's work.

The comparison of quantum mechanics and Whitehead's system should begin with Whitehead's "atoms of reality": Can we find "actual entities" in quantum mechanics? That is, which quantum mechanical concept (if any) corresponds to Whitehead's concept of an actual entity?

An actual entity has inner and outer, or subjective and objective, aspects. Subjectively, that is, for itself, it is "a throb of experience." Objectively, that is, for the rest of the universe, it is "a stubborn fact," something that did happen, and, as such, an item for prehension. Since science cannot deal with the subjective side, let us concentrate on the objective aspect. When we do that, the correspondence we seek becomes readily accessible: *The objective aspect of Whitehead's actual entities corresponds to a process of collapse that leads to the appearance of an elementary quantum event in spacetime.*

An elementary quantum event is "a flash of existence." It appears out of a background of potentialities through the atemporal process of collapse and disappears back into it almost as soon as it appears. In addition, through its momentary appearance as an actuality, it creates a discontinuity in the field of potentialities. These elementary quantum events are the fundamental building blocks of the actual physical universe.

This description of elementary quantum events is in full accord with Whitehead's description of the objective aspect of actual entities. As far as the world outside it is concerned, an actual entity makes its appearance as a unique event and disappears almost immediately. Moreover, its occurrence brings about an abrupt change in the background of potentialities out of which it appeared. That it has become "a stubborn fact" changes the possibilities for the future. The same is true for an elementary quantum event. For example, an electron that is about to hit a TV screen becomes an elementary quantum event by hitting the screen at a specific location, and, as a result, its wave of future potentialities undergoes an abrupt change and starts spreading from this specific location.

The correspondence between actual entities and elementary quantum events becomes even more striking when we consider the question of identity. As we saw in Chapter 13, Section 3, elementary particles do not keep a thread of identity in time. Let us summarize our findings by quoting Schrödinger:

We are now obliged to assert that the ultimate constituents of matter have no "sameness" at all. When you observe a particle of a certain type, say an electron, now and here, this is to be regarded in principle as an *isolated event*. Even if you do observe a similar particle a very short time later at a spot very near to the first, and even if you have every reason to assume a *causal connection* between the first and the second observation, there is no true unambiguous meaning in the assertion that it is the same particle you have observed in the two cases. The circumstances may be such that they render it highly desirable and convenient to express oneself so, but it is only an abbreviation of speech; for there are other cases where the "sameness" becomes entirely meaningless; and there is no sharp boundary, no clear-cut distinction between them, there is a gradual transition over intermediate cases. And I beg to emphasize this and I beg you to believe it: It is not a question of our being able to ascertain the identity in some instances and not being able to do so in others. It is beyond doubt that the question of "sameness," of identity, really and truly has no meaning [italics in original].[19]

As we saw in Section 3, this lack of identity applies to Whitehead's actual entities in precisely the same fashion: Each actual entity is unique, separate and distinct from other actual entities. The continuous endurance of things in nature is only apparent; the impression of continuity is due to the similarity of different actual entities, and different "societies" of actual entities, as they succeed each other. In other words, there is no "entity" that stays identical from one moment to the next. We have the *impression* of identity over time simply because ever-new actual entities keep creating similar patterns.

7. A Grin Without a Cat

Having read, and reread, the first six sections of this chapter, I felt uncertain. They seemed to be clear enough to me, but would they be clear to my readers? I decided to consult with Peter and Julie. I invited both of them to have tea with me and had them read these sections. We sat together silently for a few minutes, and then Peter turned to me.

"I was impressed by what was said in our last conversation," he said, "and now, reading these ideas, I'm beginning to appreciate the possibility that the shift to a Whiteheadian paradigm can be a profound experience, but I'm not sure. How about you? What do you think? What is the real significance of this paradigm shift?"

"Well," I responded, "let's start with this: If I asked you to characterize the universe in very broad strokes, what would you say?"

"Let me see," he answered. "First of all, it's a material universe. This table, this cup of tea, your body and mine, the earth, the stars, the galaxies—they are all chunks of matter. Yes, that's what the world is."

Julie looked disappointed. "Is that all?" she asked.

Peter was puzzled. "Well, what else is there?"

"How about ideas, art, music? How about love?"

"You may not like to hear it, but I'll tell you what I think," Peter responded. "First of all, the items you mentioned have a material component. Art, for example, is patches of color on canvas; music is sound vibrations in the air; even love has the material component of excitations in the nervous system. Now, I agree that these are not the whole of what art, music, and love are. But the rest is . . . how shall I put it? Not quite real; it's all kind of fluff."

Julie was ready to lash back, but I interrupted her. "Terrific," I said. "Everything is material, and what is not matter is—how did you put it?—'kind of fluff.' But what is matter? I'm not asking for a definition, mind you. We are just trying to get at the gut-level meaning of the paradigm. How would you express, in your own way, what the word 'matter' stands for?"

Peter was mildly indignant. "If you're not asking for a definition, then we both know very well what matter is. It is this hard, impenetrable stuff that's all around us, the stuff that just sits there, doing nothing. When you kick it with your foot, as Dr. Johnson demonstrated, your foot hurts. It is the stuff that you physicists work so hard to understand."

"Yes," I said, "in fact, we worked so hard to understand it that we now understand that it does not exist at all."

"What do you mean?"

"We speak loosely about 'the electron,' but each time the presence of 'the electron' is ascertained by a measurement, it is an altogether new electron. This holds for everything that is made out of electrons, protons, and neutrons, that is, all this 'matter' all around us. This table seems to you to stay the same because certain patterns of elementary quantum events keep repeating."

"I think I follow. But can you elaborate a bit?"

"All right," I said, "suppose we are in a movie theater. We see the same Charlie Chaplin present on the screen moment after moment. But this illusion of a permanent presence is created by many separate, distinct film frames."

"Aha!" Peter exclaimed. "That's it! Matter is like the *film* in your movie analogy! *It is that substratum that carries the patterns we see all around us!*"

"That's not a bad idea." I said, "Unfortunately, it is wrong."

"How so?"

"To the best of my judgment—and I am speaking now as a physicist—this carrier of patterns you introduced simply isn't there." I pulled two books out of the bookshelf. "Schrödinger explains the situation with his customary lucidity." I found the passage I wanted and read aloud:

> Let us now return to our ultimate particles and to small organizations of particles as atoms or small molecules. The *old* idea about them was that *their* individuality was based on the identity of matter in them. . . . The *new* idea is that what is permanent in these ultimate particles or small aggregates is their shape and organization. The habit of everyday language deceives us and seems to require, whenever we hear the word 'shape' or 'form' pronounced, that it must be the shape or form of *something*, that a material substratum is required to take on a shape. Scientifically this habit goes back to Aristotle, his *causa materialis* and *causa formalis*. But when you come to the ultimate particles constituting matter, there seems to be no point in thinking of them again as consisting of some material. They are, as it were, *pure shape*, nothing but shape; what turns out again and again in successive observations is this shape, not an individual speck of material [italics in original].[20]

I turned to the other book. "Heisenberg makes a similar statement: He says that 'the smallest units of matter are, in fact, not physical objects in the ordinary sense of the word; they are forms, structures, or—in Plato's sense—Ideas, which can be unambiguously spoken of only in the language of mathematics.'"[21]

Julie was amused. "This disappearance of matter," she said, "reminds me of that scene in *Alice in Wonderland* where Alice meets the Cheshire Cat." She found the book on my bookshelf and continued: "Here we are. Alice says to the cat,

"I wish you wouldn't keep appearing and disappearing so suddenly; you make one quite giddy!"

"All right," said the Cat; and this time it vanished quite slowly, beginning with the end of the tail, and ending with the grin, which remained some time after the rest of it had gone.

"Well, I've often seen a cat without a grin," thought Alice; "but a grin without a cat! It's the most curious thing I ever saw in my life!"[22]

I nodded as she closed the book. "Yes," I concurred, "the patterns we see all around us are like the grin without the cat. The concept of an underlying material substratum is spurious. It is just a figment of our thinking; nothing corresponds to it in reality. Within customary modes of thought this absence of matter is indeed 'the most curious thing I ever saw in my life!' But the Whiteheadian paradigm does not include the assumption of an underlying material substratum. Its absence is an aspect of the paradigm. This is, incidentally, another point of agreement between Whitehead's paradigm and 'things observed.'"

Peter was puzzled. "But here is this actual world all around me. If it is not matter, what is it?"

"I'll rephrase your question: If the old paradigm is wrong, what is the new one?"

"I would like to suggest a first step toward an answer to Peter's question," Julie said. "It is this: *That which exists in actuality is that which can be experienced.*"

Julie made us both stop in our tracks. We thought about her formulation. Finally Peter remarked, with some hesitation: "It seems to me that your statement doesn't mean very much. Anything can be experienced."

"Not at all." Julie was firm.

"Okay. Give me an example of something that can't be experienced."

"Easy: A potentiality cannot be experienced. It can be thought about, but not experienced."

"Wait! Isn't thinking an experience?"

"Sure it is. But what is experienced in this instance is the thought about the potentiality, not the potentiality itself."

"All right," said Peter, "I accept that. But your definition of actuality is so abstract that it doesn't do anything for me."

"That's because you are so materialistic in your thinking!" The impatient streak in Julie's character was showing. "If you accept my definition, then the absurd limitation you imposed on yourself a few minutes ago just evaporates. This cup of tea, Mozart's symphonies, Rembrandt's paintings, my love for you—all exist in actuality."

Peter was a little distracted by the last item on Julie's list, but he kept his course: "Are you saying that all these items—this cup of tea, Mozart's symphonies, and, er, um, and so on—they are all the same?"

"I'm not saying that at all!" Julie protested. "On the contrary, I am implying a hierarchy, one that contradicts what you suggested before. In the sense of being actual, yes, all these items are the same in that they are all actual. But now your beloved material substratum is out of the way. So, if you wish to have a sense of hierarchy, it cannot be based on matter. You said that a symphony is 'fluff' compared to this cup of tea, and the whole of our culture agrees. A piece of land, rather than a great idea, is called 'real estate!' I submit that Mozart's symphonies have a much more real kind of actuality than this cup of tea."

"But why?"

"Because of my definition. I said, 'That which exists in actuality is that which can be experienced.' Compare the range and depth of experiences induced by Mozart's symphonies to those induced by this cup of tea. If you wish to judge the 'realness' or 'depth' or 'level of being' of something, you should base your judgement on the 'realness' or 'depth' or 'level' of experiences that it can induce."

Peter was impressed, and so was I. After a long silence he turned to me and said, "I'm beginning to see what you mean by a change of paradigm."

"You haven't seen anything yet," I responded. "All we have done so far is take the first small step away from materialistic realism."

8. Shifting Gears

The hallmark of a successful new paradigm is the easy resolution of problems that are, from the point of view of the old paradigm, puzzling and disturbing. Magellan's ships set sail westward and, without changing course, arrived back to their port of origin. How difficult to explain within the paradigm of a flat earth! And how easy, once the paradigm shift from a flat earth to a spherical earth has taken place!

As we have seen, some features of the physical world that were discovered by quantum mechanics are puzzling and disturbing. The puzzles appear, however, because we formulate them and think about them in Newtonian terms. If we think in Whiteheadian terms, the strangeness fades away. Let us summarize our findings so far.

Whitehead's system is a comprehensive and complex paradigm for the understanding of reality. One of its key ideas is this: Reality is "atomized" in the sense that the elements of actual existence are discrete. These discrete elements, these "actual entities," are processes of self-creation. The different phases in the development of each one of these processes do not follow each other in time. Rather, their final outcome is, from the point of view of other actual entities, the emergence of "stubborn facts" into space and time. Furthermore, these actual entities in themselves are not objects. They are, first and foremost, "throbs of experience." Experience is the primary characteristic of all that exists in actuality.

This perspective sheds light on much that seems obscure in our observations when expressed in the language of the Newtonian world-view. However, when we try to verify the new paradigm scientifically, we run into a problem: The methodology of science, by concentrating exclusively on the objectivized aspects of the entities studied, excludes the possibility of detecting the presence of experiences.

While a scientific verification of the primacy of experience is impossible, the question of whether the objective aspects of Whitehead's system are in accord with the findings of science can be examined. Such an examination reveals that Whitehead's system provides natural explanations for the findings of quantum mechanics. Features that are strange and puzzling when we follow our habitual ways of thinking become simple and natural within a Whiteheadian way of thinking. Since Whitehead's system itself is comprehensive and self-consistent, this agreement with the findings of quantum mechanics supports our confidence in Whitehead's system in general and the idea of the primacy of experience in particular.

The interplay between the potential and the actual is central to the ontological interpretation of the quantum theory. In the next chapter we will look at the relationship between the potential and the actual in some detail, both in the context of quantum mechanics and in the context of Whitehead's system, thereby deepening our understanding of both.

16. The Potential and the Actual

The transition from the potential to the actual is a central element in both Whitehead's process philosophy and quantum mechanics. This chapter is devoted to a deeper analysis of this transition. Using "delayed-choice experiments," the relationship of potentialities to time is shown to be different than the relationship of actual events to time. This leads to the question of whether a quantum state describes the potentialities themselves or the knowledge that is available about them. We end with a fresh look at Bell's experiment and at the concept of locality.

~

"Curiouser and curiouser!" cried Alice (she was so much surprised that for the moment she quite forgot how to speak good English).

—Lewis Carroll

1. The Coexistence of Contraries

According to Whitehead, each and every actual entity is a creative activity, a process of self-creation. This activity is described as "the process of eliciting into actual being factors in the universe which antecedently to that process exist only in the mode of unrealized potentialities. The process of self-creation is the transformation of the potential into the actual, and the fact of such transformation includes the immediacy of self-enjoyment."[1] Let us explore this aspect of the creative process—the transformation of the potential into the actual—within Whitehead's system and compare it to its counterpart in quantum mechanics, the collapse of quantum states.

The distinction between the potential and the actual is central to many philosophical systems, including most systems of ancient Greece, as well as Whitehead's own system, and is also central to the conceptual framework of quantum mechanics. It is appropriate, therefore, to begin our exploration with the question "What is the essential distinction between the potential and the actual?"

We use the words "actual" and "potential" every day, and we know very well what we mean. Standing under an actual oak tree, I see an acorn on the ground. The acorn is a *potential* oak tree. This not only means that it may (or may not) become an oak tree in the future, it also means that it is not an oak tree now. This is clear enough in the context of everyday life, but what is the distinction between

179

the potential and the actual in the quantum context, when these terms are applied, for example, to electrons?

In both contexts the essential distinction is the same: *Competing potentialities coexist peacefully, while competing actualities obstruct each other.*

Right now I am faced with a decision of whether to paint the walls of my room white or blue. Two potentialities are now present: the potentiality of the walls being painted white, and the potentiality of their being painted blue. These two potentialities exist peacefully together because, as potentialities, the presence of one does not obstruct the presence of the other. Tomorrow morning, however, when I actually paint the walls, only one of these potentialities will become actual. If one of them will take place, the other will not. Therefore, as actualities they are mutually exclusive.

This idea can be summarized as follows: *Potentialities are subject to the logic of "and," while actualities are subject to the logic of "either/or."* Right now the walls of my room are potentially white *and* potentially blue. Once they are painted, however, they will be *either* white *or* blue.

This distinction is as valid in the atomic and subatomic domains as it is in everyday life. Consider the impingement of an electron upon a TV screen. Before the impingement takes place, while the electron's existence is potential rather than actual, the potentialities for impingement at all points of the screen coexist peacefully. Once the impingement occurs, however, it will occur at a particular point. Actual impingement at point A and actual impingement at another point B mutually exclude each other.

2. Delayed-Choice Experiments

While the distinction we have just discussed is common to both everyday situations and the quantum domain, the one that we are about to consider was discovered for the first time in the context of quantum mechanics. It has to do with the relationship of potentialities and actualities to time and seems rather mysterious. This new distinction can be stated as follows: A change in present conditions cannot affect past actual events, but, in certain circumstances, a change in present conditions can bring about a change in the history of potentialities. Loosely speaking, it can bring about a rewriting of this history. Situations where past potentialities change to accommodate present conditions are called "delayed-choice experiments." Here is an example:

Recall the experimental set-up we discussed in Chapter II, Section 3. Figure II.I shows an arrangement of ordinary mirrors and half-silvered mirrors in which the following takes place: A beam of light arrives from the left and is split in two as it hits the half-silvered mirror on the left. The part of the beam that is reflected moves along path a, and the other part, passing through the half-silvered mirror as if it were a transparent piece of glass, moves along path b. The reflected beam is diverted by the two ordinary mirrors and then joins the transmitted beam. The joining occurs as they both hit another half-silvered mirror, the one on the right. Both beams are now split in two by this half-silvered mirror. The beam that followed path a splits into (1) a beam reflected toward the left and (2) a beam that passes straight through and continues downward. The beam that followed path b splits into (3) a beam that continues moving to the left and (4) a beam that is reflected upward.

Next we considered what happens when one turns down the intensity of the incoming beam so much that the incoming beam becomes a flow of individual photons: One photon arrives, goes the whole way, and leaves the apparatus; then another photon arrives, and then another, and so on. We can no longer speak of half-silvered mirrors splitting beams. We have to consider what happens to each individual photon as it encounters those mirrors.

Toward the end of Section 3 of Chapter 11 we studied a variant of this experiment (shown in Figure 11.2) in which one of the fully silvered mirrors, the one on the top right, is not rigidly supported; it is mounted on a light spring and will start vibrating whenever anything hits it. The spring is assumed to be so light that even a single photon can make it vibrate.

The discussion of this experiment and its variant led us to the following conclusions: In the original experiment, when all mirrors are rigidly supported, each photon will be in a state of superposition of (1) going along path a and (2) going along path b. Loosely speaking, it would be "to a certain extent" along a and "to a certain extent" along b. This way of speaking is loose, because there is no actual photon traveling along either path. The superposition is a superposition of potentialities.

In the variant of the experiment, i.e., when one of the mirrors is free to vibrate, a superposition of path a and path b is no longer possible. Loosely speaking again, each photon follows either path a or path b. The path it took can be determined experimentally: If it travels along path a, it hits the spring-mounted mirror, whereas if it travels along path b, it does not. Hence the path of travel can be determined by watching whether or not the mirror vibrates. Once again, this way of speaking is loose, because there is no actual photon traveling along either path. The statement "The photon travels along path a" means that there is a potentiality for detecting a photon along path a if a measurement apparatus is placed along it. The spring-mounted mirror is such a measuring apparatus. Because of its presence, the quantum state of the putative "photon traveling along path a" collapses at the spring-mounted mirror to become an actual, elementary quantum event through the act of impingement upon the mirror.

To summarize, let us compare the experimental arrangements depicted in Figures 11.1 and 11.2. Loosely speaking, in Figure 11.1 the photon is *not* "choosing" a path. It gets through the apparatus as a superposition of potentiality waves traveling along paths a and b. In Figure 11.2 the photon *is* "choosing" a path. It "travels" along either path a or path b, the choice of a or b being made at random for each photon.

A choice is being made, then, between two possibilities: Possibility 1 is a passage of the photon through the experimental set-up *in a state of superposition of paths a and b*, and possibility 2 is a passage through the experimental set-up *through one or the other of the two paths*, either path a or path b. When is this choice being made?

The obvious answer is this: This choice is being made when the photon enters the apparatus, that is, when it meets the half-silvered mirror on the left. The factor that determines the choice is the mounting of the mirror at the top right. If it is rigidly mounted, possibility 1 prevails; if it is mounted on a spring, possibility 2 prevails.

But wait! This answer cannot be right. Consider the following scenario: When the photon enters the apparatus, the top right mirror is rigidly mounted; hence the photon's path is a superposition (possibility 1). But while the photon is en route, someone changes the mounting of the mirror and makes it spring-mounted. This is happening while the photon is past the point of choice and is already in a state of superposition! What is the poor photon going to do?

Answer: No problem. It will switch, in midair, so to speak, from possibility 1 to possibility 2. It will rewrite its history and behave as if possibility 2 were chosen when it entered the experimental set-up.

This variant of the original experiment, i.e., *the addition of a change from rigid mounting to spring mounting after the photon has entered the apparatus*, is what makes it a delayed-choice experiment.

Two comments need to be made at this point. First, the experiment we are discussing is, obviously, a thought experiment, not one that can be carried out in practice. The speed of light is so great that there is no time to change mountings during the time of passage of a photon through an apparatus. But equivalent experiments have been carried out, employing electronic switches, rather than hardware mountings, to accomplish the delay in choice.

Second, the language used in the preceding paragraphs was very loose indeed. It needs to be "translated" from descriptions of flying photons to statements of changing potentialities. Doing that, however, would make the text hopelessly opaque. Therefore I will leave the text as it stands and trust that you understand what it is supposed to mean when properly "translated."

We started this section with a statement of a new distinction between potentialities and actual events. Using the delayed-choice experiment as an example, we are now in a position to fathom its meaning. Imagine that we are doing the experiment. A photon has already entered the apparatus, and we are changing the mounting of the mirror from rigid to spring-mounted. This actual event of changing the mounting affects the history of the field of potentialities of the photon. The past of this field of potentialities changes, from a superposition (possibility 1) to a path selection (possibility 2).

It follows that the statement "An actual event is taking place now" is very different in its relation to time from the statement "The potentiality now is such and such." The first statement indicates the emergence into existence of a "stubborn fact" that will never be eradicated. Past potentialities may change, however, depending on present actual events.

3. The Potential Enters the Actual

Delayed-choice experiments indicate a curious flexibility in the relationship between potentialities and time. The reason for this flexibility is this: Unlike actual events, potentialities do not exist in spacetime; they merely *refer* to specific locations and specific times. And they do not exist in spacetime because *potentialities as such are eternal.* What does this mean?

Consider the distinction between "eternality" and "everlastingness." "Everlastingness" is the opposite of "change." It means endurance, forever, in time. By contrast, "eternality" means staying the same because the very nature of the item in question is unrelated to time. An example: If one believes, as Aristotle did, that the universe has no beginning and no end, one can say, "The universe is everlasting." This statement means that the universe could have a beginning and/or an end, but doesn't. By contrast, the number three is not everlasting; it is eternal, because it has nothing to do with time. Numbers, mathematical relationships, colors, sounds, scents, Beauty, Truth, and Justice, in short, all of Plato's Ideas, *as well as potentialities*, are eternal. Whitehead calls them "eternal objects."

When we try to understand an actual entity we have to pay attention not only to how it "appropriates" the "stubborn facts" of its universe but also to the nature of its relationship to the eternal objects. For example, the understanding of a leaf includes understanding its relationship to the shade of green it manifests, as well as to the shades of green it does not manifest: If the shade of green of a particular leaf is different from that of its neighbors, this may indicate that something is wrong with it. Similarly, the understanding of the leaf also includes understanding its relationship to shape—the shape it actually has, as well as other shapes it might have had, but doesn't. In short, the understanding of an actual entity involves a reference not only to what is but also to what could be, or, in other words, not only to actualities but also to potentialities.

The relationship of an eternal object to other eternal objects is determinate. For example, the number three is related to the number nine in that it is its square root. The relationship of an eternal object to an actual entity, or to a society of actual occasions, however, is a priori, indeterminate (Whitehead calls this relationship "an ingression"). For example, a particular painting may include three lemons, and it may include four. The mode of ingression of eternal objects into an actual occasion is a solution of this indeterminacy, whereby the relationship, which is a priori indeterminate, becomes determinate. For example, this particular painting could have included any number of lemons; when it is finished, however, this indeterminacy is resolved: It includes just three.

4. The Collapse of Quantum States

Our discussion of the relationship between eternal objects and actual entities has a direct bearing on the understanding of the collapse of quantum states. Remembering that the collapse is the counterpart, in the conceptual framework of quantum mechanics, of the objective aspect of an actual entity, we are now in a position to see how Whitehead's system explains the delayed-choice experiment.

Consider the possibility of the self-creation of a particular actual entity, such as a photon impinging on the spring-mounted mirror. The choice as to whether or not this actual entity will arise involves a choice among potentialities, which are eternal objects. These potentialities are (1) maintenance of a superposition of states along paths a and b as the photon travels through the apparatus, (2) selection of path a, and (3) selection of path b. These potentialities, as potentialities, are eternal. They coexist peacefully and do not become determinate, except when there is a possibility of the arising of an actual entity.

Such a possibility arises when the photon can impinge upon the spring-mounted mirror. Potentiality 1 is ruled out, at that moment, if the mirror is, indeed, spring-mounted. There remains the choice between potentialities 2 and 3. If potentiality 3 is chosen, no impingement takes place, and there is no actual entity. If potentiality 2 is chosen, the photon becomes actual as it impinges upon the spring-mounted mirror. The crucial point is this: *No choice among potentialities 1, 2, and 3 will be made until the time when the impingement is possible.* The first actual entity that can create itself in the experiment is the photon impinging on the spring-mounted mirror. The moment of this possibility, and not the time of the presumed entry of the photon into the apparatus, is the moment of choice. The thought "A photon is entering the apparatus" does not correspond to any actual entity. Therefore it is just a thought; it does not describe an actual event.

The phenomenon of delayed choice is strange if we think of it in terms of a material photon passing through the apparatus. It ceases to be strange if we think of it in terms of the relationship between actual entities and eternal objects.

5. Description of Reality or Description of Knowledge?

We are ready now to come back to two issues that we left unresolved. The first is the issue of the ontic versus the epistemic interpretation of the wave functions or quantum states (see Chapter 4, Section 5), and the other is the apparent faster-than-light propagation of influences in Bell's experiment (Chapter 7, Section 7). The resolution of these issues expresses my own conclusions, rather than a consensus among physicists.

How would Bohr and Heisenberg have responded to the question of the ontological status of wave functions? In Section 5 of Chapter 4 we compared their views on the question "is nature described by quantum mechanics?" Bohr's response would have been "absolutely not. Quantum mechanics does not describe nature. It describes what we can say about nature." Heisenberg's response, on the other hand, would have been "Yes, of course. Quantum mechanics tells us what atomic and subatomic particles are really like. They are fields of potentiality that become actual when measured." Well, who is right?

As we have seen in Chapter 4, Section 2, Max Born figured out the correct way of using the wave functions: the square of the wave function indicates the *probabilities* of finding the particle at different locations. Does this give us an answer to the question as to their ontic or epistemic nature?

The purpose of the present section is to find out the answer to this question. I felt that Julie's sharp intellect, as well as Peter's wish for complete understanding would be helpful at this stage of our journey.

When I explained to Julie and Peter that I wished them to engage with this question they were interested. It has become customary lately, in our discussions, to have Peter begin by looking for precise definitions and clarifications. Today's conversation started in the same vein.

"This question we are facing," Peter began, "sounds important. It even contains two impressive terms, 'ontic' and 'epistemic.' Would you start, Julie, by telling me what they mean?"

6. What Does "Ontic" and "Epistemic" Mean?

Julie was not inclined to go into long explanations. "'Ontic,'" she said, "means related to being and 'epistemic'—related to knowledge. What's unclear about that?"

"It is the expression 'related to' that is unclear," Peter explained. "Here is what I mean. Suppose we find some interpretation of wave functions. I want to know whether it is ontic or epistemic. Can you give me a criterion, a way to tell which it is?

Both Julie and I felt that this time Peter found one of the key questions right away. We were quiet. After a few minutes Julie spoke.

"We don't need to go to wave functions to answer your question, Peter. Let us take a simple example. Suppose a dog bites Shimon, right here. What kind of event would that be?

"Ontic, of course." Peter responded. "It is related to a real event, here and now, rather than to knowledge."

"Wait, not so fast. What if *I tell you* that a dog had bitten Shimon? You haven't even seen it. Or I may be lying to you. Furthermore, according to what I tell you, it happened far away."

"I see," Peter responded. "The same event can be ontic or epistemic, depending."

"Depending on what? Remember, we are looking for a criterion."

Peter felt as if the rug was being pulled from under him. He did not like the feeling.

"I can feel," he said, "that this little discussion of ours can easily expand to fill a philosophical treatise. It all depends on what happened as well as on who knew what. It is all very interesting, but I don't want to go there."

One could barely notice Julie's smile. "All right" she said, "I have a way out for you. Don't worry, you won't go where you don't wish to go."

"Yes?"

"We won't resolve any deep philosophical question. We will merely find out about ontic vs. epistemic *quantum measurements*. In other words, we will use the same strategy that was used by Einstein and Co. when they dealt with the EPR issue. We covered that in Chapter 6, Sections 5 and 6. In the case of EPR Einstein and his collaborators decided not to try to solve the question of what is an element of reality in all its philosophical glory. They gave up on trying to find *a sufficient* rather than *a necessary condition* for something to be an element of reality. We will use the same approach in relation to our inquiry, applying it, of course, not to elements of reality but to quantum measurements."

Peter didn't quite see the light at the end of the tunnel, but he did feel better. "O. K.," he said, what's your sufficient condition?"

Peter was not ready for Julie's response. All she said was, "location."

"What do you mean?" he blurted out.

"Look, Peter. Suppose you have an epistemic interpretation of the wave function and suppose the wave function undergoes a collapse. Where do you think the collapse takes place?"

Peter thought very hard, and then, unexpectedly, his face broke into a smile. "Of course," he said, "in the brain of the knower. Where else? That's because if the interpretation of the wave function, or quantum state, is epistemic, a collapse means a change in the field of knowledge. Knowledge requires a knower, and knowers have brains."

7. The Double Slit Pattern

"Well, Peter, not so fast." Peter felt as if Julie were getting ready to send his way one of her curve balls. "Remember how complicated the double slit pattern was, the one we studied on pp. 28 and 29 (Figures 3.4 and 3.5)? You don't really believe that such a pattern can be produced by any brain that happened to be around! Add to that the question of whether or not this brain had some martini before constructing the pattern, and the whole thing becomes absurd."

Once again Peter had to admit that Julie had a point. "O. K.," he said, a valid interpretation of a wave function cannot be epistemic. So, it must be ontic."

"That's where you are wrong. Wave function cannot possibly be ontic." Julie was merciless, and Peter was thoroughly confused. "But why?" he blurted.

"Because," Julie continued her explanation, "of the weakness Einstein discovered in the ontic interpretation (see Chapter 6, Section 2). In these early days of exploring quantum mechanics no one even considered the possibility that the epistemic interpretation is right; they took it for granted that what they were dealing with were "real" waves. And Einstein realized right away that since the collapse of the waves is simultaneous with the measurement and, furthermore, it takes place instantaneously, something will have to propagate faster than light, because instantaneous is faster than anything." And so, Einstein considered a weakness in the (implicit) ontic interpretation as a weakness of quantum mechanics."

8. Is the Collapse Instantaneous?

To Julie's surprise, Peter came up with a new argument against Julie's position. "Look here," he said, "your argument, the one you have just used to exclude the ontic interpretation, is based entirely on the belief that the collapse is instantaneous all over space. What if it isn't? What if it takes some time for the collapse to propagate from the location of the quantum measurement to other locations? If the collapse propagates at the speed of light, or lower, your argument falls flat."

Julie gave her friend a look of appreciation. "You are right," she said. "The question now becomes, is it or isn't it possible to prove that the collapse propagates faster than light? This question was taken up by two physicists, Yakir Aharonov and David Albert, who proved without the shadow of a doubt that the collapse has to be instantaneous. I won't reproduce the proof—it is too complicated for the level of our discussion here. However, I have just checked it with our friend Shimon. He assured me that the ontic interpretation cannot be saved this way."

"All right," Peter conceded. "In this case, we have to accept the epistemic interpretation." He sounded dejected. "But I must tell you that I don't like it."

"Well, you are not alone. Remember the concept of potentiality that helped us understand what an electron is (Chapter 3, Section 4)? Now it will help us settle the issue of the collapse. You, Peter, don't like introducing the physicist's brain into our interpretation, and your dislike is justified and understandable. However, what if we define the epistemic interpretation in terms of potentialities? You remember that potentialities are a category of existence, different from both existence and non-existence. So, what the physicist in charge does or doesn't do, whether or not he gets drunk, makes no difference. This is possible because of the kind of entities fields of potentialities are, because they don't exist in actuality. As we saw in Chapter 4, Section 4 potentialities can *refer* to different, even contradictory events or properties. For example, before I buy it, my car is potentially white as well as potentially black. When the potential becomes actual, it will be either white or black. A reference to an actually painted car requires that it is painted in the chosen color, black *or* white. Try to pull the same trick on actually existing entities—you will be dead in the water."

Peter was impressed. "Thanks, Julie," he responded. "So, modified epistemic interpretation it is. It is modified by replacing actually existing fields by fields of potentialities."

"That's right, Peter," Julie said, "you got it."

Coming back from my conversation with Julie and Peter, I was convinced that Julie was right in her belief that the epistemic interpretation is, indeed, the right one, and that the quantum state, or wave function, describes not the actual knowledge but the potential one: if certain data about quantum systems are available, the quantum state cannot possibly be affected by whether or not someone chooses to read that data.

Siding with the potential epistemic interpretation against the ontic one is not, however, the end of our story. It turns out that whenever the quantum system under consideration is simple as well as non-relativistic (that is, when the effects of Special Relativity, such as the relativity of simultaneity, are hardly relevant) *the two interpretations converge*. This comes about as follows.

Consider a simple quantum system, such as an electron. An electron can be subjected to a complete set of measurements, following which its quantum states is knowable. This quantum state includes all of the information that is available about it—the probabilities all of the possible results of all possible measurements; that is, it includes the full contents of the field of potentialities that we call "the electron." Now consider this: The electron exists as an actual entity only when measured. Between measurements the electron is a field of potentialities the full contents of which are encapsulated in the quantum state. It follows that *the description of what is knowable about it, i.e., its quantum state, is identical with the description of what it is.*

In Whitehead's paradigm, the description of what an actual entity is cannot be identical with a description of what can be known about it through measurements and observations. Every actual entity is an experience, and the experiential aspect is available only to itself. But an electron that is not being measured is not an actual entity. It is, rather, the set of potentialities for the results of all possible measurements; and these can be known. For a non-relativistic electron, Bohr's interpretation and Heisenberg's interpretation prove to be identical!

The identity of the two interpretations is valid for simple systems but not for complex ones. Our argument is based on the assumption that it is possible to carry out a complete set of measurements on the system under consideration. This is not true, however, for complex systems.

To summarize: In principle, if by the word "epistemic" we refer to available rather than actual knowledge, then the epistemic interpretation is the correct one. In most situations that are encountered in practice, however, the relativity of simultaneity does not play a role, and the quantum systems under consideration are relatively simple. When these conditions prevail, the two interpretations converge, and we are free to adopt whichever we prefer.

9. A Fresh Look at Bell's Experiment

The relationship between quantum mechanics and Whitehead's philosophy is a two-way street. Quantum mechanics adds credence to the Whiteheadian vision, and Whitehead's system helps us understand the apparently weird features of the quantum domain. We saw an example of this relationship in our discussion of "elementary quantum events" as the objectivized aspect of Whitehead's "actual entities" (Chapter 15, Section 6). In the present section we will reexamine Bell's correlations from a Whiteheadian perspective. This reexamination will aslo lead to a further clarification of the concept of "locality."

We left the discussion of the violations of Bell's correlations in an unsettled state (Chapter 7, Section 7): These correlations seem to show that *something*, which we called "influence," travels faster than light, but this faster-than-light travel cannot be harnessed to transmit *signals*. Apparently, influences can travel faster than light, while signals cannot. How can we endeavor to understand this strange state of affairs?

In an EPR or Bell's experiment two events that take place at the same time seem to influence each other, regardless of the distance between them. In principle this distance can be astronomical. Even when the events take place very far apart they seem to be "entangled"; they seem to "feel" each other. It has been suggested that such a connection takes place because *both events form a single creative act, a single "actual entity," arising out of a common field of potentialities.*[2] A single act of transition from the potential to the actual that occurs in two places is not the result of the propagation of anything between these two places; hence the speed-of-light barrier does not apply. This is why such creative acts cannot be utilized to transmit signals faster than light: Each experimenter is in control of some of the *conditions* to which this act of self-creation is subject. He can determine which component of the spin is measured by his apparatus, but the result of the creative act is not completely in his hands: he cannot determine whether the spin measured by the apparatus in his spaceship will end up being "up" or "down."

To transmit and receive a signal, two creative acts are required. The transmission and reception of a signal is, precisely, the creation of a situation where one completed actual entity determines another. Such transmissions cannot propagate faster than light. The distinction between influences and signals reflects the distinction between two events that are components of a single act of self-creation and two events that are connected, yet distinct, creative acts.

A mundane analogy may help clarify this distinction: Think of a dancer in the act of performing. The single creative act we have been discussing corresponds to the dancer gracefully lifting her leg and right arm in one harmonious movement. In contrast, the two connected but distinct events correspond to the appearance of an itch on her left leg, to which she responds by scratching it with her right hand. The graceful lifting of hand and leg is really a single movement; its two correlated components take place simultaneously. The two movements of itching and scratching are separated in time, however, since one is a reaction to the other. The simultaneous movement of hand and leg is an expression of a single mental act, whereas the feeling of itch and the urge to scratch are two distinct mental events.

In Chapter 8, as we began our exploration of the consequences of Bell's discovery regarding the paradigm of local realism, we stated that, regardless of whether or not we give up realism, locality has to go (Section 2). Now we are in a position to sharpen this statement. When we say that the tenet of locality cannot be saved, what exactly do we mean by "locality"?

The assumption of locality says: "Nothing can propagate faster than light." But in the case of a propagation of an influence from point A to point B, the influence is not transmitted by a physical entity that departs from A and arrives at B. Rather, a measurement that is carried out at point A changes the potentialities that affect the creation of a single actual entity that is localized at both points A and B, and this change is not subject to the speed-of-light barrier.

This state of affairs leads to the conclusion that the statement "Nothing can propagate faster than light" can be given two different interpretations: (1) No physical entity propagates faster than light; and (2) nothing that happens at point A can affect what takes place at point B faster than the time it takes light to travel from A to B. It is the second meaning of the term "locality" that is violated in the EPR and Bell experiments: "locality" as interpreted in the first statement is not violated.

10. The Nature of the Whiteheadian Paradigm

The last two chapters were devoted largely to the introduction of Whitehead's world-view with a possible modification. We found that if process philosophy is to account naturally for the apparent non-local propagation of "influences" in Bell's correlations it needs to be modified. The suggested modification was compatible with the main thrust of Whitehead's world-view.

There remains the task of examining whether process philosophy, either in its original or modified version, is (1) realistic, (2) materialistic and (3) deterministic.

(1) There is no doubt that process philosophy is realistic. Actual entities are assumed to exist "from their own side," independently of human consciousness.

(2) It is equally clear that materialism is out. Indeed, the characterization of actual entities as "throbs of experience" shows that they have an experiential aspect. According to Whitehead the universe we live in is a universe of experiences rather than a universe of objects.

(3) Since the Whiteheadian universe is a creative one, and since determinism, or "a clockwork universe" is the very antithesis of creativity, there is little doubt that according to process philosophy our universe is anything but deterministic.

PART THREE

Physics and the One

Whitehead's process philosophy provides a metaphysical foundation for the understanding of reality. It leaves, however, essential questions unanswered: Does reality consist of levels, some of which are "higher" than others in a profound sense? Do human beings have a place, and a role to play, in the cosmological scheme?

These questions are outside the domain of Western science in general and quantum mechanics in particular. And yet, surprisingly, the mysterious "collapse of quantum states" continues to be a rich source of suggestions. The collapse, the process of transition from the potential to the actual, involves a selection: There are many possibilities, only one of which is actualized. How is the selection made? According to Paul Dirac, "Nature makes a choice." Can we arrive at a concept of nature as an entity that makes choices?

In the third century A.D. the neoplatonic sage Plotinus presented a multileveled concept of reality that includes the idea that nature contemplates, and produces by contemplation. Such an idea seems strange to the modern mind. However, a close analysis of the laws of nature in general, and the conjunction of order and randomness in particular, shows that the discrepancy between a system such a Plotinus's and the quantum theory is not as wide as it seems.

According to Plotinus's vision, reality is multileveled, its higher levels are reached through inner journeys into the depth of the Subject, and, furthermore, we have a significant place in the cosmological scheme. The basic limitation of our science, i.e., its adherence to the principle of objectivation, prevents the quantum theory from being in correspondence with Plotinus's vision. Once again, however, a look at quantum measurements in general, and at the role of the observer in particular, suggests intriguing ideas.

We end Part III with a chapter entitled "Physics and the One." The central quest of present-day physics is the search for "a theory of everything," a set of equations that contain, in principle, solutions to any and all problems. This outer quest is compared with the millennia-old inner quest, the quest for what Plotinus called "the One." We examine the subject/object mode of perception and the possibility of its transcendence and conclude that as long as the quest of physics excludes the subject of cognizance, it is bound to remain incomplete. True unification must be based on a transcendence of the principle of objectivation.

17. Levels of Being

To complete our sketch of the emergent world-view we need to go beyond Whitehead and tackle questions that he left unanswered. In this chapter we inquire whether the universe has a hierarchical structure. We show that the existence of different levels of experience points to the existence of different levels of being. Such a hierarchical structure is an integral part of Plato's and Plotinus's philosophies. How do these philosophies compare with the findings of physics? When we remember that physics is subject to the principle of objectivation, similarities emerge: First, the laws of nature can be recognized as objectivized Platonic Forms; second, the collapse of quantum states can be interpreted in terms of the noumenal world's participation in the phenomenal.

~

Many times it has happened: lifted out of the body into myself; becoming external to all other things and self-encentured; beholding a marvelous beauty; then, more than ever, assured of community with the loftiest order....

—Plotinus

1. Levels of Experience

The idea that reality is made up of "throbs of experience" is the cornerstone of Whitehead's philosophy. Whitehead developed a comprehensive theory of the general characteristics, structure, and genesis of these "throbs of experience," or actual entities. He paid little attention, however, to the hierarchy among them. The only hierarchical idea that Whitehead dwells upon is the difference in level between God, who is, like any other element of reality, a "throbs of experience," and other "throb of experience." Generally, however, the hierarchy among actual entities is not an important element in process philosophy.

As we have seen in the previous chapter, knowledge with utter certainty, while possible, is not easy to come by. In Heisenberg's case, such a high level of knowledge came in a moment of correspondingly high level of experience, one that transcends the subject/object mode. To deepen our understanding of levels of knowledge, we turn now to the exploration of these levels.

For the exploration of the levels of experience and levels of being we will turn to Plato, the fountainhead of Western philosophy, and to his remarkable disciple in spirit, Plotinus. After we review some relevant aspects of their thinking, we will turn back to contemporary physics and compare its findings with the Platonic paradigm. We will end this chapter with some comments about the relationship between the ideas of Plato and Plotinus and the ideas of Whitehead.

Plato exerted a considerable formative influence on Heisenberg, and Schrödinger was so immersed in Greek philosophy that he wrote a book about it, *Nature and the Greeks*. This interest of the founders of quantum mechanics in Greek philosophy was not accidental.

Before we get to Plato and Plotinus, however, let us look at our own experiences. Our scientific bias does not encourage us to classify our experiences in terms of levels, to consider some as being "higher" or "more real" than others. Yet, when we rely on intuition and common sense, we can't help doing just that.

Consider, for example, Heisenberg's experience at Prunn Castle, when he felt connected with "a central order" (Chapter 10, Section 3). Isn't it clear that such an experience is on a different level from my experience of reading the newspaper? Isn't an experience of being deeply moved by a work of art, by a piece of music, by a child's suffering, different in quality from experiences of daydreaming while waiting for the bus?

This point is so obvious that it does not need elaborating. We can easily categorize any of our experiences as being more or less "deep," more or less "real," along a continuum, or dimension, of "realness." This dimension is quite different from the dimensions of space and time. In space and time there is no essential difference between myself, standing in front of the Mona Lisa and being deeply moved, and myself, standing and daydreaming while waiting for the bus.

The exploration of experiencing is the theme of Marcel Proust's celebrated set of seven novels having the joint title *Remembrance of Things Past*. At the climax of the seventh novel, *Time Regained*, Proust explores a sequence of inner events in which past moments are experienced in the present not as memories but as if he is reliving them.

Proust describes himself walking, in a dejected mood, to a party:

> Revolving the gloomy thoughts which I had just recorded, I had entered the courtyard of the Guermantes mansion and in my absent-minded state I had failed to see a car which was coming towards me; the chauffeur gave a shout and I just had time to step out of the way, but as I moved sharply backwards I tripped upon the uneven paving-stones in front of the coach-house. And at the moment when, recovering my balance, I put my foot on a stone that was slightly lower than its neighbor, all my discouragement vanished and in its place was the same happiness which at various epochs in my life had been given to me by the sight of trees which I had thought I recognized in the course of a drive near Balbec, or by the sight of the twin steeples of Martinville, or by the flavor of a madeleine dipped in tea, and by all those other sensations of which I have spoken.[1]

Proust goes on to describe the intense experiences that followed, as well as his search for the source of the sudden happiness and of the vision behind the transformatory experience triggered by his stepping on the uneven paving-stones.

And almost at once I recognized the vision: it was Venice, of which my efforts to describe it and the supposed snapshots taken by my memory had never told me anything, but which the sensation which I had once experienced as I stood upon two uneven stones in the baptistery of St. Mark's had, recurring a moment ago, restored to me complete with all the other sensations linked on that day to this particular sensation, all of which have been waiting in their place—from which with imperious suddenness a chance happening had caused them to emerge—in the series of forgotten days.[2]

As Proust continues to examine the experience, especially the happiness and freedom from all fears, including the fear of death, which characterized it, he discovers that the being who enjoyed the experience was not his ordinary self.

The truth surely was that the being which had enjoyed these impressions had enjoyed them because they had in them something that was common to a day long past and to the present, because in some way they were extra-temporal, and this being made its appearance only when, through one of these identifications of the present with the past, it was likely to find itself in the one and only medium in which it could exist and enjoy the essence of things, that is to say: outside time. This explained why it was that my anxiety on the subject of my death had ceased at the moment when I had unconsciously recognized the taste of the little madeleine, since the being which at that moment I had been was an extra-temporal being and therefore unalarmed by the vicissitudes of the future.[3]

Proust's account of his extraordinary experience was followed by a clear expression of the understanding that as one moves into a depth or a height of experience, one moves *away* from time and into an extra-temporal dimension. As we will see, this example can help connect some of Plato's claims, which are formulated in his Dialogues in an abstract language, with concrete human experiences.

2. The Sources of Experiences

In the previous section we discussed the range of the levels of *human* experience. According to Whitehead, however, *all* elements of reality are "throbs of experience." If human experiences exhibit many levels, how many more levels are needed to characterize all the visible and invisible aspects of reality—all of which are experiences! How are we going to encompass this enormous range? How do we compare the experience of a worm wriggling in dirt with Mozart's experience of writing his Requiem?

We need a principle of comparison—and it seems that Julie, in her definition of the actual, provided us with a clue. As you may recall from Section 7 of Chapter 15, Julie responded to Peter's question about the nature of the actual by saying,

"That which exists in actuality is that which can be experienced." According to this line of thinking, Mozart's experience is superior to the worm's to the extent that it can induce in other sentient beings—us, for example—more profound experiences. *The suggestion is to judge the level of an actual entity by its potentiality to induce deep experiences.* While this approach does not yield a quantitative measure, it does suggest a reasonable way of thinking about the levels of actual entities.

Julie's approach leads, however, to a curious puzzle: Certainly, a great work of art has a greater potentiality to bring about deep experiences than a kitchen chair. However, looking at one of van Gogh's masterpieces depicting a kitchen chair, it is clear that a kitchen chair did evoke a deep experience in him. Similarly, Proust's transformatory experience was triggered by the simple sensation of walking on uneven paving-stones. How are we to reconcile the general validity of Julie's approach with such remarkable exceptions?

We must invoke the distinction between *levels of experience* and *levels of being*. We will explore the idea of levels of being in the succeeding sections, but, for the moment, we will begin the discussion by considering only two levels of being, as they are discussed by Plato: the realm of Being (the noumenal world) and the realm of becoming (the sensible world), or, in other words, the Intelligible realm of eternal Forms and the realm of transient phenomena. In every experience both realms are represented, but the degree of participation of the noumenal varies: For example, the more beautiful an object, the more the Form of Beauty is present in it. From this point of view the level of an experience can be judged by the extent to which the presence of the noumenal in it is not veiled by the transient aspects.

This way of thinking explains why great works of art are "higher" or "more real" than the act of reading a newspaper: The presence of the Form of Beauty is more transparent in them. Heisenberg's experience in Prunn Castle was "high" because he experienced the Form of Order more or less directly.

However, if we recognize that the noumenal is present in every experience, every object, and every event (opaque as this presence may sometimes be), then the paradox we have just raised in relation to van Gogh is resolved: A great artist is gifted with the capacity to experience the noumenal in the mundane and to render the mundane on a piece of canvas in such a way that some degree of that experience is evoked in the viewer. The degree of evocation depends, of course, not only on the artist's work but also on the viewer's openness to it. Proust's experience is similarly accounted for. Not only was he a very sensitive human being, but the experience of the "extra-temporal," i.e., the noumenal, in the mundane came to him when he was ripe for it; in some sense the whole sequence of seven novels is the account of how he became ripe for the experience.

3. The Experience of the Noumenal

Plato's world-view is encapsulated in a wonderful story known as the cave allegory, told in the seventh book of *The Republic*.

The cave allegory presents a vision of reality that consists of three major levels of being: first, "the Good," the highest and the most real, the source of the being of the next level; next, the Intelligible realm of the many Forms (other than the Good) which eternally are; and last, the sensible world of transient phenomena in space and time, phenomena that are the shadows of the Forms. These transient phenomena are "shadows" because they do not have an independent existence; the source of their existence is the being of the Forms. We, who are conditioned to be informed merely by our senses, mistakenly consider the sensible world to be independently existing and the only reality there is. We are in this respect like the prisoners in the cave, who mistake the shadows for the objects that cast the shadows.

Is it possible for us to be disabused of our error and experience the true reality, first the Forms, and eventually the Good? A close reading of the cave allegory indicates an affirmative answer. Plato gives a specific example of someone who has to argue in the courts of law after having seen "Justice itself." To be sure, such a person is not commonly met with. The access to the real world is not easily won; it requires "a steep and rugged ascent." It is possible, though, and one cannot help feeling that Plato himself, as well as his teacher Socrates, was not a stranger to this world.

Plato was not in the habit of invoking his experiences, or Socrates', as evidence. At the conclusion of the cave allegory he even inserts a disclaimer. "You will not misapprehend me," Socrates tells Glaucon, "if you interpret the journey upwards to be the ascent of the soul into the intellectual world according to my poor belief, which, at your desire, I have expressed—whether rightly or wrongly God knows."[4] When we come to Plotinus, however, the experiential account is explicit:

Many times it has happened: lifted out of the body into myself; becoming external to all other things and self-encentered; beholding a marvelous beauty; then, more than ever, assured of community with the loftiest order; enacting the noblest life, acquiring identity with the divine; stationing within It by having attained that activity; poised above whatsoever within the Intellectual is less than the Supreme; yet, there comes the moment of descent from intellection to reasoning. . . .[5]

If the actual is that which can be experienced, we are compelled to consider the noumenal world as actual. All accounts of the experience of it, from Plotinus in the third century A.D., to Proust and Heisenberg in the twentieth, agree on one point: When it is experienced, it is experienced as more real than the sensible world; ordinary experiences pale in comparison, to the point of becoming devoid of meaning.

Plato's vision of three levels of being is our starting point in the discussion of the hierarchy of what is. We turn now to Plotinus for a fuller account of this hierarchy, an account that is based on both his personal experience and his formidable intellectual capacity.

4. The Hierarchy of Being According to Plotinus

In broad outline Plotinus introduces five levels of being: the One, the Nous, Soul, nature, and the sensible world.

The One—the counterpart of Plato's "the Good"—is the supreme principle of unity, which defies thinking. It can be pointed to, but not grasped, by thought; and even this "pointing" is of questionable validity. The One is beyond thought because thought, by its very nature, is discriminating: Thinking about X implies a distinction between X and not-X. When I think about "the One," I make a distinction between "One" and "not One." I am thinking of two items; my mind is not in accord with the One itself. Similarly, even the distinction between "being" and "non-being" does not apply to the One. Its transcendence is so complete that one cannot even say, "The One is."

The Nous, on the other hand, is the principle of being. It *is* in the fullest sense of the word "is." And it is a characteristic of being itself that it is not different from knowledge; moreover, *its knowledge is knowledge by identity.* "The Nous possesses, or rather is, truth, because in it there is an identity of knower, knowing and being known, of contemplator, contemplation and object of contemplation."[6]

Plotinus's doctrine of Soul, the next level in the hierarchy, is rather complex. Essentially Soul is Nous. It is, however, that aspect of Nous that is oriented toward matter. The term "Nous" is used in two different meanings depending on the context: as Nous entire or as the higher aspect of Nous, in contradistinction to Soul and nature. Soul itself has a myriad of aspects: the World Soul, which is oriented toward the world; individual people's souls, oriented toward individuals; souls of animals and plants. Just as Soul is an aspect of Nous, these different souls are not distinct; they are not other than Soul itself: They are merely its different facets.

In the fifth *Ennead* Plotinus explains the difference between the Nous's and the Soul's modes of knowing:

Its [Nous's] knowing is not by search but by possession, its blessedness inherent, not acquired; for all belongs to it eternally and it holds the authentic Eternity imitated by Time which, circling around the Soul, makes towards the new thing and passes by the old. Soul deals with thing after thing—now Socrates; now a horse: always some one entity from among beings—but the Intellectual-Principle [i.e., Nous] is all and therefore its entire content is simultaneously present in that identity; this is pure being in eternal actuality; nowhere is there any future, for every then is now; nor is there any past, for nothing there has ever ceased to be; every thing has taken its stand for ever, an identity well pleased, we might say, to be as it is; and everything, in that entire content, is Intellectual-Principle and Authentic Existence.[7]

Soul, like the Nous, is atemporal: time is "circling around" it. And yet, unlike the Nous, "Soul deals with thing after thing—now Socrates; now a horse." The Nous contemplates through the Form of Identity, while Soul contemplates through the Form of Difference. Hence we can conceive of Soul's contemplation as separable into distinct acts or elements.

Nature is the lower part of the World Soul:

[Nature] constitutes a level lower than that of soul proper, an eternally fixed level of contemplation which falls short of the clear contemplation of the blessed souls which remain in the Nous. Yet, at least according to the entire "gradation" side of Plotinism, matter could not be formed without this lower part of the World Soul.[8]

Finally, the sensible world is like a reflection of the noumenal world in matter; the sensible world has no being of its own. It is the product of nature's contemplation, as a dream is a product of the dreamer's mind, and *matter is the principle that makes such a reflection possible.* Now, just as a reflection of myself in a mirror is merely an appearance of myself and does not contain my "being," the sensible world, qua sensible, has no being. Since matter is that principle which makes the appearance of such "non-being" possible, matter *is not.* It is "the last of the forms" and has "the being of non-being."[9]

The image of the sensible world as a reflection of the noumenal is merely a metaphor. While revealing one aspect of the relationship between them, it conceals another. In the case of reflections in a mirror there is a spatial separation between the object and its reflection; the object is in front of the mirror, while the reflected image seems to be behind it. No such separation exists between the sensible world and Nous. *Nous is the being of the sensible world.* Time and space distinctions are not applicable to the Nous, because, while the being of anything may be *associated* with appearances in time and space, being, qua being, is not in space and time. For example, my own being is associated with my body but, as being, is not anywhere in space and time.

When I first encountered Plotinus's system, it felt strange, arbitrary, and abstract, verging on the meaningless. As I became more acquainted with it, however, it grew on me. Its profundity, subtlety, and truth became more apparent. The experience of becoming acquainted with Plotinus's thought became akin to the slow acquisition of the capacity to read poetry in a foreign language, to begin to receive what such poetry can convey. In particular, I discovered the need to be constantly on guard against the unconscious translation of Plotinus's expressions into my ingrained scientific paradigm. We will come back to this point in Section 6, when we discuss the relationship between Plotinus's idea of Nous and our idea of the laws of nature.

5. Nature Contemplates

Playing at first, before we handle our subject seriously, if we were to say that all things desire contemplation and look to this end, not only rational but also irrational animals, and that nature in plants and the earth which engenders them, and that all gain it, each one having it according to the nature of each, but that different ones gain contemplation in different ways, some truly, while others get an imitation and image thereof—would anyone sustain so paradoxical a statement?[10]

Plotinus is careful in the way he introduces the idea that nature contemplates, and produces by contemplation. The idea must have been as strange for his contemporaries as it is for us, and yet, as the continuation of *Ennead* III.8 shows, he really means it.

An idea feels strange if it contradicts our paradigm. It may be instructive, therefore, to check out such an idea with the poets, who are relatively free of the paradigm of their times. Because poetry is an expression of direct experience, poets make paradigms transparent by the way they use them to transcend them. Recall Wordsworth's expression of his experience of the "presences of Nature":

> Ye Presences of Nature in the sky
> And on the earth! Ye visions of the hills!
> And Souls of lonely places![11]

In Shelley's *Prometheus Unbound*, which, like Wordsworth's poem, is quoted by Whitehead in his *Science in the Modern World*, the earth speaks as a being that is quite capable of contemplation:

> "I spin beneath my pyramid of night,
> Which points into the heavens,—dreaming delight,
> Murmuring victorious joy in my enchanted sleep;
> As a youth lulled in love-dreams faintly sighing,
> Under the shadow of his beauty lying,
> Which round his rest a watch of light and warmth doth keep."[12]

Such poetic expressions can help bridge the gap between the idea that nature contemplates and our scientific conditioning, which treats rivers and mountains as inanimate. Plotinus, however, provides us with much more than a poetic expression of an experience. His idea that nature contemplates and, through contemplation, produces, is an integral part of a fully articulated world-view.

Truth, Being, Knowledge, Life, and Contemplation are different ways of naming what the Nous is. The conceptual distinctions among these terms do not correspond to real distinctions in the Nous. They correspond, rather, to distinctions in *our* experience, to the different aspects that may appear in our experience of the Nous, of Being itself, when this experience is less than complete.

The level of the Nous is so high that most oppositions are reconciled in it. For example, in our usual mode of knowing, I, the knower, am distinct from, and stand over against, *it*, that which is known. In the Nous, however, knowledge is by identity: Knower, known, and knowledge are one. Perhaps we approach such an experience while being with another person and having, for an instant, a feeling of deep communion.

Another opposition that is reconciled in the Nous is the opposition between activity and passivity. Again, the possibility of such reconciliation can be indicated at extraordinary moments. Dancers and athletes sometimes speak of "the still point" from which their movements originate when they are at their best. Moments of intense contemplation can be described as moments of deep quiet that are the very opposite of the static and the inert.

As we saw in Section 5 of Chapter 10, contemplation is very different from discursive reasoning. Contemplation has to do with contacting the timeless dimension of the noumenal realm. Hence it is not so strange to conceive of the highest form of contemplation as the activity of the Nous itself, an eternally intense activity that is not different from an eternal quietude. And, as Soul and nature are aspects of the Nous, they too share this activity of contemplation, albeit on lower levels.

The preceding discussion may have shed some light on the idea that nature contemplates, but how to understand the second half of Plotinus's idea, namely, that nature *produces* by contemplation? This idea too is an integral part of his world-view: *Each level of being is produced by the one above it through contemplation, which is not different from mere presence.* The producer does not intend to produce; nor does it make efforts. Its mere presence is enough—and this way of production is production at its best. Thus the One produces Nous, Nous produces Soul, Soul produces nature, and nature produces the sensible world. With the sensible world, production comes to an end, because the sensible world, qua sensible, *is not*. It has no being and is like a reflection of being; naturally, a reflection is not capable of producing anything.

Not only is production through contemplation unintentional and effortless, it does not detract anything from the producer. Consider, for example, the "production" of an image in a mirror. The image does not take away any of the substance of the reflected object. Likewise, an object that casts a shadow does not use up any of its substance in casting that shadow. Similarly, the production of Nous from the One, or the production of Soul from Nous, or nature from soul—all these productions are a kind of "overflowing" of the presence of the producer, with one important difference: Unlike ordinary, physical overflowing, the activity of producing does not diminish the producer in any way.

Our discussion gives rise to at least three questions. First, if the production occurs through the producer's mere presence, why introduce the idea that the producer produces by contemplation?

Take, for example, the production of nature by the World Soul. That which is produced is highly ordered; hence it is a manifestation of Intelligence. The idea of contemplation conveys a hint as to the kind of intelligent, eternal activity that the World Soul is engaged in. It produces by its mere presence *as a contemplator.*

Second, if the producer contemplates, what does it contemplate? Does it contemplate its product?

Not at all. The Nous contemplates itself, World Soul contemplates the Nous, and Nature contemplates World Soul. In the case of the World Soul, for example, it is the contemplation of the perfect intelligence and order of the Nous that gives rise, as a kind of an unintentional overflow, to the order of nature. Perhaps we can liken the World Soul to a poet who contemplates the stars in rapt, wordless ecstasy. The poem he eventually writes, wonderful as it may be, is an afterthought that pales in comparison with the wonderment of his experience.

Third, ordinarily the idea of production is associated not with contemplation but with a physical doing, a physical activity that follows a plan or a decision. How can we understand a purely mental activity, i.e., contemplation, as productive?

This issue goes to the heart of Plotinus's vision. Its elucidation calls for the convergence of a number of ideas. First, our ordinary idea of production is based on viewing the sensible world as "substantial." For Plotinus, however, the sensible world is like a reflection or a dream. Just as we produce dreams by a purely mental activity, Nature produces the sensible world by a higher kind of dreaming: "We are such stuff as dreams are made on."

The same principle holds for the other levels. Unlike the sensible world, which is *not*, nature *is*. The level of its being, however, is so much lower than Soul's that Soul's contemplation is enough to produce it as a kind of an unintentional aftereffect. A similar principle applies to the production of Soul from Nous and of Nous from the One.

Second, even in ordinary doing, the visible, physical activity is merely the tip of the iceberg, especially when the production is a skillful one. The physical movements of a master craftsman are imbued with mental concentration, intention, the mostly unconscious activity of making use of previous experiences, and so on.

Third, if we wish to understand why the images we see in a mirror move the way they do, we need to explore the movements of the real objects in front of the mirror, the objects whose reflections we are watching. We will never understand the causes of the movements of the reflected images by analyzing the images as if they existed by themselves, independently of the objects. Similarly, since for Plotinus the sensible world is merely a reflection of the noumenal, whatever we see happening all around us in the sensible world, including physical doing, is really due to the being-knowledge, or contemplation, of the Nous, and even beyond that—to the presence of the One.

Finally, in the highest form of human production, such as the production of an artistic or musical masterpiece, real contemplation seems to be the essential ingredient. Artists reporting about peak moments of creativity sometimes speak of a purely spontaneous, contemplative kind of activity, from which planning, correcting, checking, etc., are entirely absent.

6. The Laws of Nature

When I think about noumenal entities, such as the Nous, Soul, or nature, I encounter a difficulty similar to the one I encounter while thinking about Whitehead's actual occasions. My mind plays a trick on me: It takes these entities and presents them as mental objects. I see the Nous, for example, as an amorphous "something" in front of my mind's eye, and yet I know that whatever Nous is, it is not that. Nous is not a "something." It is Being, unobjectivized and unobjectivizable.

When Plotinus describes his ascent to the Intelligible realm, he begins by saying, "lifted out of the body into myself; becoming external to all other things and self-encentered." The first step in the ascent is a turning away from the objectivized, moving into the unobjectivizable world of self.

Ordinary mind has this tendency to start the process of thinking by making the item that we wish to think about a mental object we can look at. This is understandable. Ordinary mind was not designed to deal with deep questions; in daily life this tendency to objectivize works very well. When we try to think about deep questions, however, we need to be careful.

It is not easy to think about the noumenal realm. Plotinus's assurance in his writing about it is born of his experience. If we are to make real contact with his ideas, we need to connect them to our own experience.

While few of us have had experiences such as Plotinus's, most of us have had extraordinary experiences of one kind or another: periods or instants of contemplation, a perfect moment while dancing or playing a musical instrument, a perfect phrase that comes to us while writing poetry. Such moments carry the taste of being in touch with a higher level, a level that transcends the mundane. In thinking about Nous, Soul, or nature, it may be useful to remember such experiences.

For example: When we think abstractly, it may be puzzling to think about the Nous as being alive. After all, isn't the Nous a kind of abstract principle, the principle of being? Not at all, Plotinus would answer. The Nous is Being itself, that is, perfect being, which implies perfect beauty, perfect goodness, perfect life. It is more than alive; it is, as it were, Life itself. "There is, then, a scale of knowledge and truth which is the same as the scale of life. The clearer knowledges, the clearer lives are the truer knowledges and the truer lives. The clearest and truest is the Nous itself."[13]

This idea that the Nous is alive may be puzzling when we think abstractly; but in the context of extraordinary experiences, like Proust's, the puzzle disappears: During such experiences one is closer to the noumenal realm; one feels, and is, very much alive. When such experiences are over, the transition back to the ordinary level of living feels like a descent. It is *as if the degree of our aliveness at any moment is the degree of our relation to the noumenal realm at that moment.* From the vantage point of such experiences it is natural to think about the Nous as the Source of Life, or as Life itself.

It was obvious to the Greeks that a higher level of being is associated with a higher level of life. In the Middle Ages, when Platonic and/or Aristotelian thinking was integrated with Christianity, the same general world-view was maintained, albeit in a different garb. God became the highest, most alive being, and the intermediate levels between Him and us were filled with a host of heavenly beings, such as angels and archangels. Newton, whose work eventually led to the collapse of this world-view, subscribed to it fully.

During the Age of Enlightenment, however, everything changed. When the great physicist and astronomer Pierre Simon de Laplace presented his book *Exposition du systéme du monde* to Napoleon, the emperor examined the book and asked where God was in relation to the scheme of nature presented in it. "Sire," Laplace replied, "I had no need for this hypothesis." We can hear the modern-day scientist speaking, precision, arrogance, and all.

For Plotinus, the sensible world is a mere reflection of the noumenal; for science, the sensible is the real thing, the only reality there is. The world of science is the world as given by the senses. Oddly, however, the Nous is not entirely absent from it. The Nous did lose its life, but not its presence. The idea that the sensible world is sustained and directed by an invisible substratum is the fundamental premise of science. This invisible substratum is called "the laws of nature."

The laws of nature are unifying principles that regulate the events of the physical world. Take, for example, Newton's law of gravity. It is the invisible factor behind the visible movement of the moon around the earth, or the earth around the sun. It is also the invisible factor behind the falling of apples from apple trees. It is, therefore, a *unifying* principle: There is nothing in the world of appearances to indicate a commonality between the falling down of apples and the movement of the moon around the earth, and yet, on the invisible level of the laws of nature, these two phenomena are closely related. In fact, they are manifestations of the same law. Similarly, Maxwell's equations of electromagnetism account for the characteristics of light, as well as for the characteristics of X-rays, heat waves, and radio transmission. As phenomena, light, X-rays, heat radiation, and radio are very different; yet they are all electromagnetic waves, obeying the same set of equations.

We believe in the existence of invisible unifying principles behind the apparent variety in the phenomenal world, and so did the Greeks. Unlike the Greeks, however, we consider these principles lifeless. For us they are mathematical relations imposed on the behavior of objects, much like the laws of the land that are imposed on the behavior of people, except that, unlike people, objects have no choice but to comply.

The fact that we understand the laws of nature as lifeless is no accident. It is a necessary consequence of what Schrödinger called "the principle of objectivation," the tacit assumption that present-day science is based on. Let us recall:

By this [i.e., by "the principle of objectivation"] I mean what is also frequently called the "hypothesis of the real world" around us. I maintain that it amounts to a certain simplification which we adopt in order to master the infinitely intricate problem of nature. Without being aware of it and without being rigorously systematic about it, we exclude the Subject of Cognizance from the domain of nature that we endeavor to understand. We step with our own person back into the part of an onlooker who does not belong to the world, which by this very procedure becomes on objective world.[14]

The removal of the Subject of Cognizance, that is, ourselves, from "the domain of nature that we endeavor to understand" is the removal of life from nature. Since Soul, the innermost part of oneself, is an aspect of the Nous, the approach to the Intelligible realm requires that one proceeds, like Plotinus, by "becoming external to all other things and self-encentered." Such an approach, however, is excluded by the principle of objectivation. A scientist, in his work as scientist, is supposed to treat his or her subject matter as objects. He or she is barred from the experience of approaching the Nous and experiencing it as the source of the phenomena of nature. This exclusion is tacit and, therefore all the more firm. But the Nous *is*. Plotinus and others have experienced it in its full glory, and most of us have intimations of it in our very best moments. It did stay on, therefore, as the basis of scientific explanations, in the form of the laws of nature. It stayed, though, within science in the only form science is capable of giving it, namely, as a corpse.

7. The Interface Between the Noumenal and the Phenomenal Realms

We make a distinction between the phenomenal (or sensible) and the noumenal for the sake of thinking. In reality they are one: *The noumenal is the being of the phenomenal*, just as the object in front of a mirror contains the being of its reflection in the mirror, and the being of the shadows in Plato's cave is contained in the objects that cast the shadows. Likewise, we make a distinction between the laws of nature and the events of nature, but in reality these are merely two aspects of the processes of nature.

To the extent that we think of either pair, either noumena and phenomena or laws of nature and events of nature, as distinct, we cannot help wondering about the relationship between the distinct items. We are faced here with the question of the relationship between being and becoming, or between eternity and time, in either the Greek or the scientific context.

I was surprised to discover that an essential aspect of the question can be addressed simultaneously in both contexts. In both paradigms, the timeless provides the time-bound, i.e., the sensible world, with a set of potentialities. The potentialities as such are timeless. The choice as to which of them will be actualized in the sensible world is the transition from the timeless to the time-bound.

Pursuing the idea that phenomena are like mirror reflections of the noumena, consider reflections of ordinary objects. The object can have many reflections, depending on where it is placed in relation to the mirror. Also, there are different reflections depending on the shape of the mirror, which can be flat, convex, or concave. Each possible reflection reveals an aspect of the object, and no reflection reveals the object in toto. Moreover, the many possibilities of different reflections are eternally present, but if we have only one mirror, and the position of objects relative to it is fixed, these possibilities can only be actualized one at a time.

We can think of electrons as quantum fields of potentialities in a similar fashion. As an electron is approaching a TV screen, all the potentialities for its impingement upon different points of the screen are present. There is, however, a moment of choice, which we called "the collapse of the quantum state." The impingement occurs at one location only and becomes an elementary quantum event.

As we probe deeper into this idea that the actualization of potentialities is the interface between the eternal and the time-bound, the parallels between Plotinus's view and the view introduced by quantum physics become truly surprising. First, as we saw in Section 4, for Plotinus matter *is not*; it has "the being of non-being." This is in full agreement with Schrödinger's statement:

> The habit of everyday language deceives us and seems to require, whenever we hear the word "shape" or "form" pronounced, that it must be the shape or form of something, that a material substratum is required to take on a shape. . . . But when you come to the ultimate particles constituting matter, there seems to be no point in thinking of them again as consisting of some material. They are, as it were, pure shape, nothing but shape.[15]

Second, and even more remarkable, for Plotinus matter is not only that which is not, it is also the expression of pure chance, pure randomness, the absence of order. Order comes from the Intelligible realm; the order that is found in the sensible world is due to the presence of the Nous; matter, "the last of the Forms," is the negation of Intelligibility and the negation of order. The framework of quantum mechanics is in full accord with this perspective. The appearance of an elementary quantum event as a result of a collapse is ruled by both randomness and order. The events are random, but, as we saw in Chapter 4, Section 2, the randomness is subject to probability distributions. These distributions are fully ordered. They are determined precisely by wave equations, such as Schrödinger's equation. Since wave equations are expressions of laws of nature, they are perfectly ordered and fully determined, as befits the noumenal realm. When applied to actual situations, the equations impose order on these situations by making their randomness follow probabilistic patterns. But a probability distribution does not eliminate randomness; it merely imposes a statistical pattern on it. The actual events are never fully ordered, because they are the meeting places of the perfect order of the noumenal realm with the pure randomness inherent in matter. And by "matter" we do not mean "a material substratum"; rather, we mean that which is not, but appears to be, as the condition for the appearance of elementary quantum events.

Finally, consider Heisenberg's view of elementary particles. Heisenberg was impressed by the fact that elementary particles are not "little things" and that their fullest description is given not in words but in mathematical symbols and equations. He concluded that "the smallest units of matter are, in fact, not physical objects in the ordinary sense of the word; they are forms, structures, or-in Plato's sense-Ideas, which can be unambiguously spoken of only in the language of mathematics."[16]

Our investigation of the meaning of these mathematical symbols tends to sup port Heisenberg's claim, albeit inconclusively. The discussion of the epistemic versus the ontic interpretation of quantum mechanics (Chapter 16, Section 5-8) led to the conclusion that under most conditions the quantum states of elementary particles describe what they are, as well as the available knowledge about them, because the two, that is, the available knowledge and their being, are the same. But, as we saw in Section 5 of the present chapter, the identity of knowledge and known is a characteristic of the Nous! For Plotinus, this identity includes the knower as well. We cannot expect the knower to show up in a scientific theory, however, because, by the principle of objectivation, science excludes "the Subject of Cognizance [i.e., the knower] from the domain of nature that we endeavor to understand."

Unlike the other parallels between the Platonic world-view and quantum mechanics that were drawn in the present section, this last parallel, suggestive and intriguing as it may be, remains incomplete: In Plotinus's view, the identity ofknown, knower, and knowledge in the Nous is always true. The convergence of the epistemic and ontic interpretations of quantum states, however, is valid in some cases and invalid in others. There is a fundamental difference between them, and when this difference is significant, the epistemic view is the correct one. This discrepancy between Plotinus's philosophy and the quantum theory deserves further study.

8. The Conjunction of Order and Randomness

Throughout this narrative we entertained three conceptual frameworks: quantum mechanics, Whitehead's, and Plotinus's. As we saw in Chapter 15, Whitehead's system is uniquely appropriate as a comprehensive metaphysical system within which the apparently surprising features of quantum mechanics find natural explanations, and we have just discussed a possible convergence between Plotinus's system and quantum mechanics.

There are many similarities between Plotinus's and Whitehead's views of reality, but there are also profound differences. In particular, for Plotinus the Forms actually exist; for Whitehead they are "eternal objects," which do not exist in actuality. Also, Plotinus's emphases are different from Whitehead's. For example, he expounds at length on the different levels of being and the relationships among them, a subject Whitehead hardly touches upon.

A full-fledged comparison between Plotinus's and Whitehead's views is outside the scope of this book. We will conclude this chapter, however, with a few comments that indicate possibilities of reconciliation between them. To begin with, the Platonic idea of Nous is comparable to Whitehead's idea of God, especially to what he calls "the primordial nature of God."[17] Unlike Plato and Plotinus, Whitehead postulates, in addition to His primordial nature, which is eternal, a consequent nature, which is in flux. As far as I can tell, however, this difference between the two systems of thought has no bearing on the relationship of either with the quantum theory. The main issue we need to address is the comparison between Plotinus's idea that nature contemplates, and produces by contemplation, and Whitehead's idea of actual entities as processes of self-creation.

When Whitehead claims that actual entities are processes, he does not mean processes in time; he means, rather, atemporal processes. Similarly, nature's contemplation according to Plotinus is not a temporal process. It is, rather, an expression of nature's life, which is not in time.

This last idea can be elucidated through an analogy: Just as the life of a movie star is not contained in the projected image on the screen, *life, like being, is never in space and time*. If we wish to use a spatial metaphor we have to say that life, like being, occurs in a different dimension.

The wax statues in Madame Tussaud's museum in London are so well made that they look alive. As displays in space and time, these statues are almost identical to their live counterparts. One can imagine, for example, Queen Elizabeth standing next to her wax copy; if she is motionless one may hesitate, for an instant, in deciding which is the real queen. The *life* of Queen Elizabeth, however, is not in the spatio-temporal appearance. Spatio-temporal appearances are mere displays. With the advent of new technologies one can easily imagine the construction of an automaton that imitates, almost to perfection, both the appearance and the movements of a living person, yet is still just an automaton.

The understanding that life, like being, is never in space and time is common to Plotinus and Whitehead. Whitehead's atemporal process of self-creation ends in a completed creation, which, when objectivized, appears in space and time. Likewise, nature's contemplation, which is atemporal, results in the sensible spatio-temporal world.

It seems to me that we can connect the Plotonic and Whiteheadian perspectives on the issue of the creation of the sensible world by *looking at actual entities as aspects of nature's contemplation*. The contemplation of nature is obviously an immensely intricate affair, involving a myriad of potentialities. If we look at the big picture and neglect important differences, we can look at actual entities as being, in some sense, the basic elements of this vast contemplative process.

We are in position now to come back to Paul Dirac's statement about the collapse of quantum states (Chapter 11, Section 2): "Nature makes a choice." The ideas we have been exploring in this book are probably far from Dirac's perspective (though not far from Schrödinger's or Heisenberg's). Nevertheless, they do provide a way of thinking about nature that incorporates his statement. A choice needs to be made, because, as Plotinus says, and quantum mechanics concurs, matter—that is, the condition of spatio-temporal appearance—involves pure randomness. The choice is based on nature's contemplation, which brings the Intelligible order to bear on matter, through the imposition of probability distributions. Each actual entity is a meeting place of the Intelligible and the phenomenal, of order and randomness. The bringing of a lawful probability distribution to bear on an actual entity's process of self-creation is an aspect of nature's contemplation. Subjecting itself to the factor of randomness, i.e., actualizing itself in a particular, randomly determined way, each actual entity is also an expression of the unlawfulness inherent in matter. An actual entity's process of self-creation is, therefore, a conjunction between nature's contemplation and matter, which is the condition of spatio-temporal manifestation. Dirac's concept of nature as that which makes choices corresponds to the idea of this conjunction.

Watercolor painters sometimes use a technique called "spattering": The artist dips a brush in paint, then flicks the paint off the brush to create a mess of tiny irregular spots of paint on the canvas. The artist regulates the spatter by the choice of paint, by the choice of the distance of the brush from the canvas as the brush is flicked, by the way his or her wrist is moving while flicking, and so on. The artist is not interested, however, in controlling the precise location of each tiny spot of paint; his or her aim is to create an overall impression of a certain kind. The randomness of the impingement of the individual spots of paints upon the paper is necessary, in fact, for the creation of the desired impression.

Nature seems to deal with the sensible world like a spattering artist. It imposes, through contemplation, an overall distribution of probabilities, and then, submitting to the randomness inherent in the existence of matter, it makes random choices regarding individual events. This conjunction of order and randomness gives rise to that distribution of elementary quantum events that does, in fact, occur.

9. Plotinus and the Emergent World-View

Let us close this chapter by raising the following question: Is Plotinus's system really the capstone of the emergent world-view?

I am attracted to Plotinus's philosophy for a number of reasons. First, its adherence to the idea of levels of being. Second, the intriguing correspondence between Dirac's view of the collapse of quantum states, i.e., "Nature makes a choice," and Plotinus's idea that nature contemplates, and produces by contemplation. Third, the fact that Plotinus's system is based not only on profound thinking but also on no less profound experiencing. And, fourth, its Platonic spirit: Since this spirit permeates Whitehead's process philosophy, Plotinus's fully articulated neoplatonic system seems to be the right one to complement it.

Nevertheless, as we saw in Chapter 14, it is the ineffable experience that comes with "utter certainty." Any conceptual formulation may prove inadequate, appropriate in one context and not in another, or even plainly wrong. We will come back to the issue of the place of conceptual systems in the Epilogue.

Whitehead, referring to his own system, expresses his view of this issue as follows:

> There remains the final reflection, how shallow, puny and imperfect are efforts to sound the depths in the nature of things. In philosophical discussion, the merest hint of dogmatic certainty as to finality of statement is an exhibition of folly."[18]

Reading the last five chapters, one may discern the outline of the emerging paradigm of the nature of reality. One vital question, however, has not yet been addressed: What is the place of humanity, what is the place of each of us, in the order of nature? This question is the subject of the next chapter.

18. Our Place in the Universe

Do we have a place in the universe? A role to play in the cosmological scheme? If the universe is largely inanimate, clearly the answer is no. If, however, it is alive and multileveled, our consciousness, especially our innate potential to reach high levels of consciousness, may well have a cosmological function. Analyses of the observer's role in the context of quantum measurements and of Heisenberg's discovery of the uncertainty principle suggest that we do have a function in the universe: to bring about a relationship between the phenomenal and noumenal levels.

~

We comprehend the wondrousness of Being, but are abashed at the thought, affirming only at rare moments that real possibilities exist. We forget that man was made to be a meeting place of the phenomenal and noumenal worlds. Every man's true nature is one of tranquility, bliss and happiness. The fact that we do not become what we are destined to be does not diminish this truth.

—William Segal

1. Do We Have a Place in the Universe?

In the vast recesses of space billions upon billions of galaxies, in clusters and superclusters, approach and, for the most part, recede from each other. Most of these galaxies are too faint for us to see, except through powerful telescopes. Only one galaxy features prominently in the night sky: the Milky Way, the galaxy we inhabit. Each galaxy contains billions upon billions of stars. Our sun, an average-size star, is situated rather far from the center of the Milky Way. Still, it is a thousand times bulkier than our very own planet earth, which circles around it as it circles around the center of our galaxy.

A billion is a very large number. Any one galaxy is an insignificant speck of dust among the billions of galaxies. Any one star is an insignificant speck of dust among the starry inhabitants of a galaxy. The earth is very small compared to the sun. We are very small compared to the earth. In view of all that, isn't it obvious that the place of human beings in the cosmological scheme is one of total and complete insignificance?

Let us examine the tacit assumptions behind this assessment. To begin with, it is based on considerations of size. Not only we, but even the earth, the sun, and the whole Milky Way are minuscule on the scale of the entire universe. Well, is size a proper measure of significance?

A billion is, indeed, a very large number. Yet, in many instances, one great scientist, philosopher, artist, explorer, political leader, or religious leader has affected the lives of billions of peoples. On the scale of the human body, the functioning of very small glands can destroy or heal a person. One virus can start a process that leads to death or to healing.

Next, consider the whole earth: We have learned in the past few decades that our actions can affect large-scale features of the planet. One or two scientists can have a significant effect on the whole earth. Consider, for example, Mario Molina and Sherry Rowland. Their discovery of the effect of man-made chemicals on the ozone layer may well have a profound effect on the future of the whole earth! One great leader can bring about a revolution in our ecological attitudes and, again, have a marked effect on the future of the planet.

The last two examples can be combined. If one virus, penetrating the body of such a scientist or leader, had a marked effect on his or her health, this one virus could affect the fate of the earth!

There is an important difference, however, between the potential effects of people on their society, or on the planet, and the potential effects of viruses on the human body: The influence of the great benefactors of humankind is due to their intentional and conscious work. As a result, this influence tends to be in the direction intended, at least in the short term and sometimes in the long term as well. The effects of Copernicus's work, for example, are still with us, still in the direction he intended—the demise of the geocentric world-view. The long-term effects of Gandhi's work are more difficult to evaluate, but in both cases, the fact that each of them has influenced millions, even billions of people is not accidental. It is due to their outstanding capacities and efforts in their respective domains of thought and action.

The case of a virus is different. As best we can tell, the effects of its actions are unintended; the consequences of its activities for the person in whose body it is lodged are accidental, as far as the virus is concerned. Indeed, in the vast majority of cases the actions of a single virus are inconsequential; only in very exceptional circumstances is the condition of a human being so precarious that the activities of a single virus tilt the balance toward sickness or health.

In general, then, *it is the presence of consciousness that makes it possible for relatively small entities, such as a single person, to affect comparatively large entities, such as a society, a civilization, or the planet.*

The sense of insignificance that modern science has engendered in us is due precisely to its lack of appreciation for the unique capacities of human consciousness. As we will see, this lack of appreciation is a feature of the paradigm, and not of the cosmos.

The cosmological picture whose outline I sketched at the beginning of this section is a recent development in human thought. The very concept of a galaxy is less than a century old. If we go back in time a mere seven centuries, we encounter the magnificent cosmological vision of Dante's *Divine Comedy*, a vision that is rather different from our own: The earth is at the very center of the universe; Rome and Jerusalem are situated at cosmologically significant locations; the span that takes the author/pilgrim from the center of the earth to the starry world, through the mountain of Purgatory, also takes him from the place of Lucifer and the eternally damned souls to the place of God, the angels, and the eternally blessed souls. There is no question that human beings occupy a central place in this universe. We are of such concern to God, the creator of the whole universe, that He sends His own Son to redeem us!

Whatever the spiritual and poetical merits of Dante's vision, it cannot be seriously entertained now as a literal representation of the physical universe. But then, will our current vision, the one I sketched at the beginning of this section, be taken any more seriously seven centuries hence? Can't we anticipate with certainty that a new cosmological paradigm will replace the current one, and isn't it possible that the issue of our place in the universe will look quite different within the framework of this new paradigm?

While the world-view that will be dominant around A.D. 2700 is impossible to foresee, it is not impossible to see important ways in which the current paradigm is flawed. It views the universe as essentially inanimate, and it replaces the variety of levels of being by one distinction: existence versus non-existence. As we have just seen, however, it is the relatively high level of consciousness that makes it possible for small entities to affect large ones. In a paradigm of an animate, intelligent universe this is hardly surprising. If we consider the universe to be alive and intelligent, the pertinent question is not the size of our bodies relative to the size of the universe but the place of our actual and potential intelligence relative to the universal intelligence, the Nous.

In regard to human intelligence it is crucial to emphasize the distinction between the actual and the potential. Clearly, the actual functioning of our intelligence, which is based largely on accidental mental associations and is concerned, for the most part, with our bodies and our egos, is, indeed, insignificant on a cosmological scale. If Plato's cave allegory does indicate a deep truth, however, the potential of our intelligence belongs to quite a different level. As long as we keep mistaking shadows for realities, our thinking is worthless, but we do have the potential to leave the cave, ascend, and be in touch with the intelligence of the Nous. If we expand our thinking to include this potential, the question of our place in the universe would appear in a new light.

The potential transition, in Plato's allegory, from prisoners in a cave to people who are free to walk in the light of the sun indicates the possibility of a change in the level of consciousness, an awakening to the noumenal world as the real one and as the being of the phenomenal. The difference between a cave dweller and a free person can be indicated by the difference between a person who is asleep and dreaming and a person who is awake. And our experiences as dreamers provide an illustration of how and why the smallness of our bodies on the scale of the universe may well be irrelevant to the question of our place in it.

Imagine yourself in bed, asleep and dreaming. In your dream you see galaxies, stars, populated planets. You see yourself as a tiny person on one small planet that circles around a star. And then, lo and behold! You see a large star advancing through this vast space toward your planet, threatening to collide with it and break it to pieces. You try to do something to save the planet, but, being a tiny person on the surface of a small planet, what can you do? You wake up in a sweat and heave a sigh of relief: Thank God this was only a dream!

While asleep, you are not aware of the fact that in reality you are not the dream figure, a tiny person on a small planet; rather, the whole dream takes place in your mind. You are the creator of the dream, and, as such, you have the potential power to change it in any way you please. Your feelings of insignificance are due to your mistaken belief that you are the dream figure, rather than the dreamer.

Our conviction that we are insignificant on a cosmic scale may well be the conviction of cave dwellers—a conviction we came to by examining only the phenomenal world around us. We may discover, however, with Shakespeare, that "we are such stuff as dreams are made on" and that the question of our place in the universe needs to be examined anew, not from the point of view of the relationship of our physical bodies to the body of the universe but from the point of view of our place in relation to the noumenal and phenomenal worlds.

While Dante's vision is obsolete in its literal aspect, it is very much alive as a poetic or symbolic representation of the human condition. Twentieth-century rational, materialistic beings that we are, it still speaks to us. And the current cosmological world-view, which may well be an *aspect* of the true picture, is certainly not the whole of it. If we try, in the spirit of Niels Bohr, to take the scientific, philosophical, and poetical expressions as complementary rather than contradictory, we may come to a fresh insight about our place in the universe.

As we saw in the previous chapters, an open-minded appraisal of the many-faceted human experience favors a view of the universe as animate, intelligent, and multileveled in the sense of being. If the universe is indeed alive, its integrity is functional: It is an ordered whole, not a random collection of unrelated clusters or superclusters of inanimate galaxies. In a living organism every part has a function. It follows that if nature is alive, the question is what our organic place in it is, and not whether we have one.

Does physics in general, and quantum mechanics in particular, have anything to say in regard to this important issue? Unexpectedly, the answer is yes. As we will see, an examination of the role of the observer in quantum mechanics suggests a paradigm shift regarding the question of our place in the universe.

Before we move on to the discussion of the role of the observer according to quantum mechanics, however, let us end the present section with an important distinction: "Having a place in the universe" may or may not involve "having a cosmological significance." The organic nature of the universe guarantees that we have a place, or a function. The function may turn out, however, to be of a local, rather than global, significance. If we compare the universe to a human body, our place may be comparable to that of a gland, the proper functioning of which is critical to the health of the body, or it may be comparable to that of a small part of a cell. In the latter case the effect of the state of that individual cell on the body as a whole may well be negligible; only the joint effect of millions of such cells is noticeable.

We will come back to the question of the possible cosmological scope of our functioning in the last section of this chapter.

2. The Role of the Observer According to Quantum Mechanics

We met John Archibald Wheeler in Section 7 of Chapter 6, when I recounted the story of his attempt to convert Einstein into a believer in quantum mechanics as a fundamental theory of nature. Wheeler has a knack for pithy formulations and simple stories that explain difficult issues. A few years ago he came up with a story that clearly enunciates the essential features of quantum measurements. The story presents a peculiar variant on the social game of twenty questions.

In the usual version of the game you are sent out of the room while the others decide on a word. When you are called back you try, as best you can, to discover the word by asking twenty questions; only yes and no answers to your questions are allowed.

Wheeler's peculiar version proceeds differently. While you are waiting outside, the others do not choose a word; instead they conspire to act as follows: As you move around the circle asking your questions, each of them is free to respond by a yes or a no, as they wish, but—and this is the key point—the responder must have in mind a word that corresponds not only to his or her answer but to all the previous answers as well.

Wheeler points out a number of parallels between the two versions of the game of twenty questions and the two concepts of measurement, the classical and quantum concepts.

First, we thought the word already existed "out there" as physics once thought that the position and momentum of the electron existed "out there," independent of any act of observation. Second, in actuality the information about the word was brought into being step by step through the questions we raised, as the information about the electron is brought into being, step by step, by the experiments that the observer chooses to make. Third, if we had chosen to ask different questions we would have ended up with a different word—as the experimenter would have ended up with a different story for the doings of the electron if he had measured different quantities, or the same quantities in a different order.[1]

Furthermore, the questioner's decision as to which question to ask determines, together with all the previous questions, the type of answer he or she will get. The questioner has no power, however, to decide whether the answer will be yes or no. Similarly, the type of experiment that a quantum system, such as an electron, is subjected to will determine the type of property it will have, e.g., whether the electron will have a definite value of position or momentum; but the experimenter has no power to determine the value that the measurement will yield. And, finally, the required consistency of the answers corresponds to the consistency of the results of different measurements.

Wheeler is quick to point out that this analogy is limited in its scope; there are important differences between the two versions of the game of twenty questions on the one hand and classical and quantum measurements on the other. Nevertheless, he concludes:

> The comparison between the world of quantum observations and the game of twenty questions misses much, but it makes the vital central point. In the world of quantum physics, *no elementary [quantum] phenomenon is a phenomenon until it is an observed phenomenon.* In the surprise version of the game no word is a word until that word is promoted to reality by the choice of questions asked and the answers given [italics in original].[2]

The subject of quantum measurements featured prominently in the previous chapters. If you go back to our discussions of it, bearing Wheeler's story in mind, you will probably enjoy some new insights. In the context of the present chapter we will concentrate on the issue of the observer's role in quantum measurements.

As we have just seen, in the game of twenty questions the questioner's role is to choose and ask the questions. The questioner has no influence over the answers. Similarly, in the context of quantum measurements the experimenter or observer can set the experiments up any way he or she wishes. Once the observer has made a decision as to the type of experiment to be performed, he or she has no influence over the outcome. In view of the fact that a quantum system, such as an electron, is, in itself, pure potentiality, it follows that the observer's role is the intentional creation of specific conditions for a transition from potentiality to actuality. The conditions determine the type of actuality that will emerge, but not its specific characteristics.

3. Physical Laws As Indications of General Truths

What (if anything) can we learn from all this about our place in the universe?

Physics deals with the simplest systems in nature. Moreover, these systems are treated at a high level of abstraction: Planets are regarded, in most cases, as point particles, muscles are considered inanimate pieces of solid material, and liquids are treated, in most cases, as if they were continuous—their discrete atomic structure is ignored. Yet the insights of physics have often served as suggestive of principles in fields of knowledge whose subject matter is far more complex, such as psychology or sociology. Why?

The reason seems to stem from a deep truth about the nature of reality: *There are fundamental principles, and types of relationship among entities, that are the same on all levels, from the simplest so-called inanimate systems to the universe as a whole.* While some principles are specific to particular fields of study, others are truly universal, applying to objects, people, societies, planets, galaxies. These universal principles and modes of relationship are easiest to discern in the simplest systems. They reach their most precise formulation, and the clearest delineation of their area of application, in the context of physics. Once understood within this context, their meaning and applications in other contexts can be explored.

Consider, for example, the attraction of opposites. In physics this principle manifests as Coulomb's law of electricity, the law of attraction between positively and negatively charged particles. This law is represented by a mathematical formula; the strength of the force of attraction can be calculated precisely, and the conditions of applicability of the law are well understood. But the principle of the attraction of opposites manifests outside of physics as well. It shows up in biology, for example, as the attraction between the sexes. Here, however, there is no simple way to quantify the force of attraction or account for its variations.

The law of inertia provides another example. As a law of physics, it states that objects stay in the same state of motion, that is, they keep moving with the same speed, and in the same direction, unless acted on by a force. Inertia, however, is a factor that opposes change in many contexts, from my own laziness to the excessively long lives of political and social institutions.

A third example, Niels Bohr's framework of complementarity, is particularly rich. Complementarity is not a principle; it is, rather, a framework for comprehending relationships between seemingly contradictory aspects of systems. Bohr discovered it as a result of his efforts to solve a problem in physics: the apparent contradiction between the particle-like and wave-like behavior of quantum systems. At the same time that Bohr discovered the framework of complementarity, Heisenberg discovered his uncertainty principle. The latter proved to be one precise formulation of the idea of complementarity in the context of atomic physics. Bohr, however, saw his discovery as applicable not only in physics but also in psychology, sociology, and epistemology. As we saw in Section 6 of Chapter 12, it applies even to the analysis of jokes!

Such manifestations of the same general principles, or frameworks of thought, in diverse fields raise the possibility that the insight obtained in the last section about the observer's role in quantum mechanics can be suggestive of a principle of universal significance. Just as the idea of complementarity opened up a new way of thinking about many issues that are unrelated to physics, the observer's role in quantum mechanics can open up an unexpected way of thinking about our place in the universe.

4. The Anatomy of Discovery

In Chapter 14, Section 2, we discussed the process that led Heisenberg to the discovery of the uncertainty principle. A reexamination of the process will reveal an intriguing similarity between the anatomy of this discovery and the observer's role in quantum mechanics.

We pointed out that Heisenberg's story is a beautiful illustration of the relationship between the unobjectivized experience of insight and its projection onto the time-bound paradigm of the culture: *The source of knowledge is a numinous experience, an intuitive insight that is incommensurable with the functioning of the ordinary, dualistic mind, the mind of contrasts and distinctions. The formulation comes about when this lower, dualistic mind comes in touch with the intuition, and does its best to express the intuition in dualistic terms, i.e., in words and symbols.* We also pointed out that in the case of science such a formulation is valid only if it is expressed in the language of science and can be checked by scientific methods.

Let us analyze this process of discovery and formulation in terms of the relationship between the noumenal and the phenomenal worlds. The uncertainty principle, an aspect of the idea of complementarity, exists in the noumenal world as an Idea, or a Form. In the moment of breakthrough Heisenberg beheld the Idea. This experience enabled him to express it and bring it to the phenomenal level of human intercourse in words and mathematical symbols, i.e., to formulate it as a coherent expression within a time-bound cultural paradigm.

The way toward the moment of breakthrough involved a conscious direction of his attention toward both worlds, the noumenal and the phenomenal, simultaneously. Heisenberg was contemplating the phenomena that puzzled him and the old principles that failed as he was striving to discover the new principle.

We are finally in a position to draw a parallel between Heisenberg's struggle and a quantum measurement. The direction of one's conscious attention in the anatomy of discovery corresponds to the creation of the conditions of a measurement: Just as the setting up of an experiment determines the *type of result* the experiment will yield, Heisenberg's mode of attending to the noumenal and phenomenal worlds determined the *kind of discovery* he could make. It had to be the discovery of a principle in physics, rather than, say, a poem or a piece of music. The analogy tells us still more: As Heisenberg was contemplating, he had no power to determine the specific nature of the discovery he was about to make; he had no power to change the uncertainty principle and come up with a different principle. Similarly, the observer of an experiment has no power to determine the result of the experiment. If he or she measures, for example, the position of an electron, he or she cannot determine the position at which the electron will be detected.

Conscious attention, an attention that is directed toward noumena and phenomena simultaneously, is no ordinary mode of thinking. Anyone who has struggled with deep issues knows that the struggle involves at least two distinct modes of mental activity: One starts out in a verbal or symbolic mode; one considers verbally formulated ideas, and looks for a solution by deliberately examining possible relationships among ideas and symbols. As this mode fails to lead to a resolution, one eventually gets into another mode, one that feels like plunging into a non-verbal, non-symbolic dimension. This is one instance of the experience of conscious attention. It is appropriate to understand such an experience as one that relates to the noumenal world, because it is this mode of exploration that can lead to the discovery of a principle. Principles, or Ideas, or Forms, belong to the noumenal world. Hence the moment of breakthrough and discovery of a principle can be considered a moment of breakthrough into the noumenal dimension, a moment of "beholding the Idea itself."

We ended Section 2 with the conclusion that "the observer's role is *the intentional creation of specific conditions for a transition from potentiality to actuality.* The conditions determine the type of actuality that will emerge, but not its specific characteristics." Similarly, in the process of discovery, the scientist's role is to create, through his or her conscious attention, specific conditions for a transition from the noumenal to the phenomenal world, i.e., a discovery and a formulation of a principle. Once the scientist employs his or her attention in this fashion, he or she has no further influence on the discovery.

The analogy between the observer's role in quantum mechanics and the scientist's role in the process of discovery is close. For the phenomenal world, the noumenal represents a wealth of potentialities. The transition that takes place in a quantum measurement from the potential to the actual parallels the scientist's transition from the noumenal to the phenomenal as he or she comes to a discovery and formulates it. This analogy is an instance of the same general principle manifesting in two different contexts.

Interesting as this analogy may be, what does it have to do with the theme of this chapter, the question of our place in the universe? To discover our place in the universe we need to contemplate our potential, rather than actual, intelligence. It is in moments of optimum functioning, such as breakthrough and discovery, that one's potential intelligence becomes actual. As we will see in the next section, the exploration of the structure, meaning, and significance of such moments will help us clarify our cosmological function.

5. Our Place in the Natural Order

We are in a position now to approach the heart of our question: What is our place in the natural order of the universe? Let us begin by reviewing, briefly, the change in the status of the question as our paradigm shifts from a clockwork universe to an organic one.

If the universe is as classical (Newtonian) physics describes it, it functions automatically. We need not wonder about our "place," because we do not have any. The feeling that we have control over anything, even our bodies, is an illusion: Our bodies are made of atoms, and every movement of every atom is fully determined by the configurations of the atoms in the universe in the distant past.

Present-day physics tells us that the universe is not Newtonian, but it cannot tell us whether or not it is alive. This issue was examined in this book, therefore, by looking for evidence from other sources. The evidence presented in Chapters 12, 14, and 17 supports the proposition that the universe is alive, intelligent and multileveled in the sense of being. We concluded (see Section 1) that since the universe is an organism, the question is not whether we have a place in it; rather, the question is "What is our place?" In view of the discussion of the present chapter, we proceed as follows:

The optimal functioning of the whole depends on the optimal functioning of the parts. When the whole of my body is in good shape, so are all its parts. And the good condition of my body has a salutary effect on my emotions and thoughts as well. Similarly, if we are organic parts of the universe, we are fulfilling our part in the functioning of the universe when we are functioning at our very best.

The analogy of an orchestra may help clarify this point. When an orchestra is playing a symphony, each member of the orchestra fulfills his or her part when, attuned both to the whole orchestra and to his or her own specific part, he or she plays well. The felt need and desire to play one's part well, and in harmony with the whole orchestra, corresponds to the collective need (for which the conductor is usually the focus) that the symphony be performed well.

I suggest, then, that in order to find our place in the universe we need to listen to our own deepest yearning and examine the moments of our most complete fulfillment. And, although the nature of our deepest yearning may not be known to us, there is plenty of evidence that moments of breakthrough into the noumenal dimension, such as the moment of Heisenberg's discovery, bring with them a sense of total fulfillment. It is natural to suggest, therefore, that *our function in the universe is to bring about a relationship between the phenomenal and the noumenal worlds.* As we saw in the discussion of Heisenberg's breakthrough, *this requires directing one's attention toward both worlds simultaneously.*

At first blush this idea may feel both strange and puzzling: What does it mean, "to bring about a relationship between the phenomenal and the noumenal worlds"?

The feeling of strangeness is due to the fact that our thinking has been formed in the materialistic paradigm that does not recognize the idea of levels of being. Within this paradigm the idea we have just proposed concerning our place in the universe makes no sense. If we open our minds, however, to a world-view that is based on the aliveness and multileveled nature of the universe, we may assuage our feelings of strangeness and find a solution to the puzzle.

In the previous chapter we became familiar with Plotinus, who understood the universe to be both alive and multileveled. Let us turn, then, to his *Enneads*:

Treating, in the *Timaeus*, of our universe he [Plato] exalts the cosmos and entitles it "a blessed god," and holds that the Soul was given by the goodness of the Creator to the end that the total of things might be possessed of intellect, for thus intellectual it was planned to be, and thus it cannot be except through soul. There is a reason, then, why the Soul of this All should be sent into it from God: in the same way the Soul of each single one of us is sent, that the universe may be complete; it was necessary that all things in the Intellectual should be tallied by just so many forms of living creatures here in the realm of sense.

. . . Our soul, entering into association with that complete [i.e., World] soul and itself thus made perfect, "walks the lofty ranges, administering the entire cosmos," and . . . as long as it does not secede and is neither inbound to body nor held in any sort of servitude, so long it tranquilly bears its part in the governance of the All, exactly like the World-Soul itself; for in fact it suffers no hurt whatever by furnishing body with the power to well-being and to existence, since not every form of care for the inferior need wrest the providing soul from its own sure standing in the highest.

The Soul's care for the universe takes two forms: there is the supervising of the entire system, brought to order by deedless command in a kingly presidence, and there is that over the individual, implying direct action, the hand to the task, one might say, in immediate contact: in the second kind of care the agent absorbs much of the nature of its object.

. . . No doubt the task of the Soul, in its more emphatically reasoning phase, is intellection; but it must have another as well, or it would be indistinguishable from the Intellectual-Principle [the Nous]. To its quality of being intellective it adds the quality by which it attains its particular manner of being: it ceases to be an Intellectual-Principle, and has henceforth its own task, as everything must that exists in the Intellectual realm.

It looks towards its higher and has intellection; towards itself, and orders, administers, governs its lower.[3]

When Plotinus speaks of "soul" he means what we call, in contemporary terminology, "self," that is, he speaks about you and me. Plotinus is saying, then, that each of us has a twofold role to play in the administration of the universe: First, our lawful participation in the responsibility of the World Soul, which is "the supervising of the entire system," and, second, the governance of the individual.

The implication of Plotinus's view is far-reaching: *The orderliness and completion of the universe depends on each one of us!* Each one of us is an aspect of the World Soul, and "The [World] Soul was given by the goodness of the Creator to the end that the total of things might be possessed of intellect, for thus intellectual it was planned to be, and thus it cannot be except through soul." And so, "the Soul of each single one of us is sent [to the 'realm of sense'], that the universe may be complete."

How can this awesome vision be related to us, as we know ourselves? Plotinus gives us transparent hints. He calls us to be "neither inbound to body nor held in any sort of servitude." This is a call to a state in which one's attention is not absorbed in concerns relating to the body, desires, or ego. Rather, one's attention is directed toward the noumenal world, the world of the unobjectivized self. One is called, in other words, to transcend the subject/object mode and enter a state of contemplation.

Note that the tasks of participation in the governance of the universe, and of the governance of one's own functioning, call for contemplation, rather than reasoning and action. The task of bringing order to the universe cannot be accomplished through the relatively inferior faculty of discursive reasoning, which figures things out in a sequential order. It calls for, rather, being in touch with aspects of the noumenal world, such as the Form of Order itself, as Heisenberg was, for a moment, during his experience in Prunn Castle.

The mention of tranquility in this context is telling. St. Augustine, who was influenced by Plotinus, speaks, in his *Confessions*, of the experience of the noumenal versus the phenomenal as the experience of eternity versus the experience of time. He makes the point, evidently on the basis of his experience, that our ordinary mode of thinking cannot approach the understanding of eternity. When people try, he says,

> their thoughts still turn and twist upon the ebb and flow of things in past and future time. But if only their minds could be seized and held steady, they would be still for a while and, for that short moment, they would glimpse the splendor of eternity which is forever still. They would contrast it with time, which is never still, and see that it is not comparable. . . . If only men's minds could be seized and held still! They would see how eternity, in which there is neither past nor future, determines both past and future time.[4]

6. Acts of Contemplation

As we have just seen, each of us has, according to Plotinus, a twofold role to play in the administration of the universe: First, our lawful participation in the responsibility of the World Soul, which is "the supervising of the entire system," and, second, the governance of the individual. A soul that fulfills this double role "looks towards its higher and has intellection; towards itself, and orders, administers, governs its lower."

To take our place in the natural order, we are called to "govern" the phenomenal level from the position of being grounded in the stillness of eternity, in the noumenal. As Plotinus puts it, "not every form of care for the inferior need wrest the providing soul from its own sure standing in the highest." Plotinus is affirming that it is possible to be in touch with both the noumenal and phenomenal levels simultaneously, that is, to have conscious attention.

There are different levels of conscious attention. The relationship between the noumenal and phenomenal dimensions can be present, through one's attention, to a greater or lesser degree. Its highest level includes the experience of their ultimate unity. The experience itself is indescribable; however, the formulations and actions that result from it always reveal unexpected relationships among seemingly distinct, even disparate, phenomena. Such revelations of deep relationships occur in different contexts. They always arise, however, with a compelling sense of certainty. Let us look at three different examples.

The first example is Proust's account of two distinct events, well separated in time, being experienced simultaneously (see Chapter 17, Section 1). Colin Wilson, the novelist, writes about such experiences:

There are certain moments in which we suddenly become totally conscious of *the reality of other times and places*. Proust described such a moment in *Swann's Way*, as the hero takes a cake dipped in herb tea and is transported back to the reality of childhood: "I had ceased to feel mediocre, accidental, mortal." The historian Arnold Toynbee described certain moments in which events in history became totally real, as if they were happening around him here and now [italics in the original].[5]

This is an example of a revelation of a deep relationship, which sometimes, as in Proust's case, reaches the point of virtual identity, between seemingly unrelated events that are well separated in space and time.

The second example is the experience of the discovery of a scientific principle, or law. Newton's law of gravitation, for instance, reveals very different phenomena, such as the falling of an apple and the revolution of the moon around the earth, to be deeply related. Both are manifestations of the same law. Heisenberg's uncertainty principle reveals limitations on possible measurements that are the same for many different atomic phenomena.

Our third example is based on a true story, told and interpreted by Joseph Campbell, the late mythologist, in one of his conversations with Bill Moyers:

In Hawaii some four or five years ago there was an extraordinary event. . . . There is a place there called the Pali, where the trade winds from the north come rushing through a great ridge of mountains. People like to go up there to get their hair blown about or sometimes to commit suicide—you know, something like jumping off the Golden Gate Bridge.

One day, two policemen were driving up the Pali road, when they saw, just beyond the railing that keeps the cars from rolling over, a young man preparing to jump. The police car stopped, and the policeman on the right jumped out to grab the man but caught him just as he jumped, and he was himself being pulled over when the second cop arrived in time and pulled the two of them back.

Do you realize what has suddenly happened to the policeman who had given himself to death with that unknown youth? Everything in his life has dropped off—his duty to his family, his duty to his job, his duty to his own life—all of his wishes and hopes for his lifetime have just disappeared. He was about to die.

Later, a newspaper reporter asked him, "Why didn't you let go? You would have been killed." And the reported answer was, "I couldn't let go. If I had let that man go, I couldn't have lived another day of my life." How come?

Schopenhauer's answer is that such a psychological crisis represents the breakthrough of a metaphysical realization, which is that you and the other are one, that you are two aspects of the one life, and that your apparent separateness is but an effect of the way we experience forms under the conditions of space and time. Our true reality is in our identity and unity with all life. This is a metaphysical truth which may become spontaneously realized under circumstances of crisis. For it is, according to Schopenhauer, the truth of your life.[6]

These three examples reveal very different kinds of unsuspected relationships, and even unity. The discovery of a scientific principle reveals an underlying unity among natural phenomena; Proust's experience reveals an unexpected relationship between distant events in his life; Campbell's story is one of a revelation of the essential identity of all human beings. In all three cases, the illusion of Plato's cave dwellers has been shattered, at least for a moment. For that moment phenomena were experienced as shadows of another reality. *At moments like these we occupy our lawful place in the universe.*

7. The Call of the Noumenal

I had not seen Peter for a few months, and I fully expected that after he read this chapter our relationship would resume its adversarial character. I knew that he was finally coming to terms with the idea that the universe is alive, intelligent, and multileveled, but the idea that our acts of contemplation matter, that they have an impact on the welfare of the universe—I expected this to be too much for him.

I was pleasantly surprised, therefore, when, after reading the chapter, he called me up and, sounding quite friendly, suggested that we meet. He and Julie happened to be in town, he said; would I care to join them and discuss "issues of mutual interest"?

We met at our favorite coffee shop, and it did not take him long to get to the point. "I read this last chapter of yours," he said, "and I became more confused than ever. What you wrote is coherent, and it makes good sense. And yet, as I was reading it, I kept getting the nagging feeling that you couldn't be right. It was weird, but sometimes I responded to your writing with a sense of awe, whereas at other times I almost pitied you for entertaining such ideas." He was looking at me with a steady gaze, trying to find out, by inspection, which of his reactions he should trust.

"Well," I thought to myself, "I wasn't really that far off. This idea that we human beings have a cosmological function is hard for him to take after all." And yet, once again, I appreciated his honesty. Julie, however, was not so appreciative. She was afraid that he had hurt my feelings. But, rather than expressing her own feelings, she stuck to the subject of discussion.

"If what you read was coherent and made sense," she asked Peter, "how can you object to it? What more do you want?"

"I want the truth," Peter responded gravely. "While it is true that accusing our friend here of being pitiful may be going too far, it is also true that fascinating, coherently presented ideas may be well off the mark."

"Yes," I agreed, "what you say is true."

There was a lull in the conversation. We felt comfortable with each other; we could share a stillness. After a while, however, I looked at Peter and saw that something was cooking in him. He was trying so hard to formulate his thoughts that when he finally spoke, he continued to stare at his coffee.

"This silence we just shared was nice," he said. "I even felt an element of contemplation in it. Yeah, it was nice for me, but surely I was not doing anything for others, or for the earth, let alone for the universe. What is this business about the universe not only needing us, but needing us to *contemplate?*"

"Yes," Julie rejoined, "this is the part that doesn't make sense to me. In the previous chapter you introduced Plotinus's idea that nature contemplates, and its contemplation brings about the required action. Here we have the same idea in relation to people: We are called to contemplate, rather than act; or, rather, our contemplation is the action that is required to bring order into the universe and make it complete. I just don't get it."

For once, Julie was on Peter's side, and he liked that. "Suppose I could do something for the universe," he said. "Then my job would be to go ahead and do it! Save the ozone layer by organizing a demonstration against the use of CFCs, or write to my congressman about global warming. That sort of thing."

"I have no objection to this kind of activity," I said, "but if Molinas and Rowland, who discovered that CFCs destroy the ozone layer, had gone to an anti-nuclear demonstration instead of thinking about the chemistry of the atmosphere, the earth would have suffered even more than it does now."

Peter frowned in response to my statement about the suffering of the earth, but he decided to let that pass. "You have a point," he finally said, "but I'm still confused."

Julie picked up the ball. "What about my question?" she asked. "How can contemplation bring about the required action?"

I felt dejected. I thought I had responded to Julie's question in this chapter, and here were my best friends, the ones I created, and I could not get them to understand! I felt as if I were in the midst of a quantum measurement. By creating Julie and Peter, I created this conversation, but I had no control over its outcome. I could not change the story and bring them over to my side without compromising my sense of integrity as a writer. I just sat there, self-absorbed and slightly depressed. Finally, however, I lifted my eyes, and Julie's pleading look made me try again.

"Look," I said, "isn't it true that all of the benefactors of mankind, the great philosophers, scientists, artists, musicians, even the few inspired ones among the political leaders, like Abraham Lincoln, were people who could and did contemplate? Isn't each great work of philosophy, science, art, and music, isn't the Gettysburg Address a message from another level of reality, a level that can be called the noumenal level? Aren't such works connections between the noumenal and the phenomenal levels? If so, could they have come into being without their creators and discoverers *living* this connection?"

Both Julie and Peter were thinking hard now, and it was Peter who articulated the question that was in the air: "I accept what you just said," he began, "but what if Mozart had had his experience of hearing a symphony and had *not* written the score down? What if President Lincoln had composed the Gettysburg Address but never delivered it? What if all these people, whom you called 'the benefactors of mankind,' had *just contemplated?*"

I was relieved. Finally, I felt, *they had found the question!*

"Thank you, Peter," I responded, "I appreciate your question. My first response will sound like a cop-out, which it is not, and the second response will go to the heart of the matter.

"The first response is this: In general, this scenario of all these great people just contemplating could not really happen. It could happen once in a while, but not as a rule. Formulation is often a spontaneous follow-up to the experience of contemplation. Moreover, processes like painting are hardly imaginable without an ongoing relationship between seeing and putting paint on canvas. In the process of writing poetry there is a similar relationship between the experiences and the words that express them. Even in the case of President Lincoln—he did not function in a vacuum; his contemplation and his actions were intertwined and a part of the political and social context he functioned in.

"This is as far as we can go *within the paradigm that all three of us are implicitly using at this very moment.* To go to the heart of the matter, however, we need to allow for a paradigm shift, which really means a total revolution in the way we think.

"To indicate the direction of this paradigm shift, consider, once again, the dreamer analogy: You are having a violent, disturbing dream, full of conflict and suffering. As long as you are dreaming and identifying yourself as a dream figure, you can only try to fix the situation by acting as a figure within the story of the dream. Imagine, however, that you, the dreamer, found a way to *change your mind,* resolve the conflicts that were expressed by the dream, and come to a blissful, quiet state. You would not have to worry about the dream any longer. Your sleep would be serene and dreamless, and if you did dream, your dreams, being an expression, or a reflection, of your state of mind, would be pleasant and free of conflict.

"Using this analogy as a pointer, try to allow for a paradigm shift in which the world we experience all around us is recognized as a projection, or a reflection, of a universal mind, the Nous. As long as we try to fix our world by demonstrations, letters to congressmen, and so forth, we are functioning mostly within the projection, and, as you know, the results are meager at best. Violence continues to breed violence. Imagine, however, that we function within a framework in which the noumenal world, rather than the phenomenal, is real. To come to order, bliss, and beneficence in our experiences, all we need to do is allow ourselves to be in touch with the Nous, to be a reflection of *its* reality, which is order, bliss, and beneficence. To be in touch with the noumenal, one needs to allow one's attention to delve into the depth of the unobjectivized Self. Within the context of the current paradigm, this is not something we even consider trying, ever. If we did, however, we might discover that in the depth of our souls we belong to the Nous. According to Plotinus the Nous acts by contemplation, and its contemplation is creative. If we found our way to this, our lawful place, we too would share in this contemplative-creative act. Our conflicts and difficulties would be revealed as the projections or reflections of the conflicts within our souls. These in turn are the results of our estrangement from the Nous, which, being the Intelligence of the Universe, is free of conflict. If we found our way back to where we belong, a resolution of our conflicts and difficulties would naturally follow."

"What you've said is really interesting," Peter responded, "but it doesn't answer my question. You speak of the potential psychological effects of contemplation. I have no problem with that. But you claim that acts of contemplation have cosmological significance, which should include physical effects as well."

"Well, the very distinction between the physical and psychological belongs to the current paradigm, not the one I'm trying to get us into," I said, "but, if you wish, we can address the question on your terms. Acts of contemplation do have physical effects. For example, when one is around people who contemplate, people who know the depth of their being as a result of their contemplation, people who are in touch with the noumenal realm, one feels something special. The effects of Abraham Lincoln's presence, for instance, have been described in these terms: He did not need to say or do anything, his very presence had a strong effect. I felt something of the kind myself on the few occasions when I met extraordinary people, such as Paul Dirac. Dirac was a quiet man, a man of few words. He was famous, in fact, for his silences. When I spent a few days with him, and with a few other physicists, back in 1976, I felt his quiet presence whenever he was in the room. To describe this experience in physical terms one would have to say that he emanated some subtle vibrations that affected people's brains in a certain way. His contemplative way of life resulted, then, in physical effects that emanated through his presence."

"What do you mean, 'he was famous for his silences?'" Julie interjected. The conversation was intense, and we were ready for a break.

"Let me tell you the story of Dirac's meeting with E. M. Forster, the novelist," I responded. "A mutual friend had learned that Dirac was interested in meeting Forster, so he invited both to dinner at his place. When they settled down, Dirac turned to Forster and asked, 'Is it possible that there was a third man in the cave?' He was referring to the cave scene in Forster's novel *A Passage to India*. 'No,' Forster replied, 'this is quite impossible.' That was the extent of the dinner conversation."

We sat quietly, enjoying each other's presence. After a while our attention expanded to include the passersby, the heavy traffic, and the tall buildings on the other side of the wide street. All of a sudden, something clicked for Peter. He turned to me and said: "This is what I'm not getting. You can speak of Dirac's contemplative nature affecting people around him, the atmosphere in the room, and so on. But look at all of this!" He made an expansive gesture with his arm. "How can anybody's contemplation affect these buildings, this traffic, the whole city and the countryside around it, let alone the solar system or the world of galaxies? It makes no sense!"

"You should be able to handle that," I said, turning to Julie. "After all, it was you who insisted, in our last conversation, that the actual is that which can be experienced."

Julie thought for a moment, then said, turning to Peter, "It is so hard to let go of a paradigm. It's like cutting off Hydra's heads—you take care of one, and a new one pops right up. You still consider the phenomenal world as existing on its own, apart from what Schrödinger called 'the Subject of Cognizance.' But the phenomenal world does not exist on its own. The concrete fact is not its own existence; the concrete fact is, rather, the experience of it, as something that, in *our* mode of experience, *seems* to exist on its own. If you, the experiencer, are stuck in a certain mode of experiencing, the universe will be, for you, fixed and unchangeable. But this is just *your* mode of experiencing, shared by the culture and reinforced by conditioning. As we discussed before, however, for some people the universe is *experienced* as alive. When you are in a different place in your Self, that is, in the noumenal world, the universe changes. Take Plotinus's testimony: 'Many times it has happened: lifted out of the body into myself; becoming external to all other things and self-encentered; assured of community with the loftiest order; enacting the noblest life, acquiring identity with the divine; stationing within It by having attained that activity; poised above whatsoever within the Intellectual is less than the Supreme.'[7] While he was in that state, the whole physical universe, which is so real for us, must have seemed to him no more than a dream. Now—and this is the main point—if the paradigm of an alive, intelligent, multileveled universe *does* represent an important truth, one that is missed in our materialistic paradigm, then Plotinus's vision is incomparably more valid than our own. He was, during his experience, at a much more intelligent level of the universe than the level where we are, even at this moment."

"Thank you, Julie," I said. "The understanding you just expressed connects with how I see the question of our place in the universe. I see the Nous as the locus of all the potentialities of experience. But the Nous itself is eternal. It is not in its nature to connect with the phenomenal world. This is our job. If we do it, the order and beauty of the noumenal will permeate our phenomenal world. If we don't, the two worlds will be cut asunder, and the universe will not be complete. The degree of strength and truth of this connection depends on the level of our experiencing, that is, the level of our sensitivity and conscious attention."

Peter's response to this last idea caught me by surprise. "Yes," he said, "this is just like quantum measurements."

Julie's intelligent look was alive with interest. "Explain," she said.

"Well," he responded, "in a quantum measurement the observer determines the set-up of the measurement, thereby determining the type of outcome. He or she cannot decide, however, what the outcome itself will be. Analogously, we have control, at least in principle, over the level of our sensitivity and attention, that is, over the quality of ourselves as transmitters from the noumenal level to the phenomenal. We can thus determine the quality of phenomena that we will experience, but we cannot determine the precise experiences that we will have. In the quantum measurement situation the precise result is selected through an unpredictable interaction between fields of potentialities and the measuring instruments. Likewise, our precise experiences are determined by our contact with the Nous as reservoir of potentialities for experience. The outcome of this contact is, from our point of view, largely unpredictable."

It was growing late, but I did not move. Julie and Peter, being the projections of my mind, did not move either. I liked their company. Slowly, however, I let them go. They faded into the night, and I was left alone, a cold cup of coffee in front of me.

8. The Significance of Conscious Attention

We started this chapter by drawing a vivid picture of the immensity of the universe relative to the size of our bodies. Our discussion revealed, however, that this disparity in size need not translate into cosmological insignificance. The thrust of the chapter was the development and explication of the idea that the call to take our place in the universe is the call to conscious attention, an attention through which the noumenal dimension can relate to the phenomenal. We are now in position to ask, "Is this a function of universal significance? Does the welfare of the whole universe depend on us?"

I am in no position to answer this question. The following comments are written as intimations or suggestions, the results of what I take to be "whispers from the other shore," an expression that is the title of a book by my good friend from Nova Scotia, the physicist and philosopher Ravi Ravindra.

As we saw in Chapter 15, the stuff that the universe is made of is not matter, as we ordinarily think of it, but experiences and potentialities for experiences. The universe is, in other words, an immense network of experiences, and potentialities for experiences, influencing each other. The immensity is vertical as well as horizontal: There are not only many experiences; there is an immense range of levels of experience, from that of the Nous to the rudimentary "throb of experience" of a speck of dust. Within this range we do have a unique possibility. We are the only beings who can relate the noumenal to the phenomenal, by experiencing both simultaneously. This uniqueness suggests a cosmological significance. This was the gist of my last comment to Julie and Peter in the previous section.

From a customary point of view such an assertion may be puzzling. Someone's experience is his or her inner affair. Others cannot touch or see it. How, then, can it be of such a monumental significance? This customary view is based, however, on the illusion of Plato's cave dwellers, who take the shadowy world of sense perceptions to be the real world. The experience of the real world is not available to our senses, and, as a consequence, the paradigm of our culture denies its existence. But in unexpected moments of breakthrough, such as the moment Joseph Campbell commented on in relation to the story about the Hawaiian policeman (see Section 6), real knowledge, whose source is not sensory information, appears with a compelling sense of certainty. In the case of the Hawaiian policeman, this real knowledge was the knowledge of the unity of all life.

In an interrelated universe of experience, all experiences affect each other. These effects are real, unbeknownst, as they are, to our ordinary minds. Thus it may well be that the order and completion of the universe does require the element that Plotinus called "intellection," brought down, so to speak, from the noumenal level to the phenomenal. And it may well be that our conscious attention is the only channel through which it can flow.

If this is the case, then our responsibility is very profound indeed, and the immensity of our potentialities is almost inconceivable. In Plotinus's account of his experience, the one Julie quoted in Section 7, he affirms the capacity of human consciousness to reach the highest levels of the Nous, to be "poised above whatsoever within the Intellectual is less than the Supreme [i.e., the One]." He thus affirms, through his own experience, the potential capacity of human consciousness to join with the highest levels of the Intelligence of the Universe.

Alfred North Whitehead did not believe that human consciousness is significant on a cosmological scale. Yet, in his last recorded conversation, a conversation with Lucien Price that took place a few months before his death, he made the following powerful statement:

The creative principle is everywhere, in animate and so-called inanimate matter, in the ether, water, earth, human heart. But the creation is a continuing process, and "the process is itself the actuality," since no sooner do you arrive than you start on a fresh journey. Insofar as man partakes of this creative process does he partake of the divine, of God, and that participation is his immortality, reducing the question of whether his individuality survives the death of the body to the estate of an irrelevancy. His true destiny as cocreator in the universe is his dignity and his grandeur.[8]

19. Physics and the One

To move on to the next phase in its evolution, science needs to incorporate the Subject of Cognizance within its domain of inquiry. As Schrödinger explains in an essay entitled "The Oneness of Mind," there are many egos but only one Mind, one Subject of Cognizance. The quest for the One Mind calls for transcendence of the subject/object mode. Can science participate in this quest? The framework of complementarity, which is based on Bohr's recognition that "Truth lies in the abyss," may be considered a first step.

~

In seeking . . . a glimpse of the incomprehensible nature of the One, the intellect must be like a bird which in flying through the air leaves no trace; the intellect's operation must dissolve back into the stillness into which it is advancing. We must think very intensely on what the nature of the One is, and then there is a point at which you let go of everything. But do not think that the discussions of this are pointless, for the One is the only important thing you can discuss ever in your life—at any time, anywhere— there is nothing more important.

—Anthony Damiani

1. Physics and the Subject of Cognizance

In the previous chapters we drew, in broad strokes, the main features of a new paradigm regarding the nature of reality. In the context of this new world-view we discussed the nature of the actual, the life and intelligence of the universe, its levels of being, and our place in it. There remains the key question, the question that Plotinus formulated in the three words "Who are we?"[1]

To be sure, no paradigm can provide an answer to this question. The answer can only come as a profound, ineffable experience, an experience that transcends paradigms. What a world-view cannot help doing, however, explicitly or implicitly, is relate to the question, encourage or discourage its pursuit, make statements about *who we are not*. If its attitude toward the question is sympathetic, it can indicate a direction of exploration and lead us to the realization of its limitations in relation to the question.

Science relates to this question by ignoring it. Ignoring can be an active, albeit negative, form of relating, as anyone who has had the experience of being ignored by a friend or a lover knows only too well. Science ignores the question, because it lacks the tools to deal with it. This attitude resembles the attitude of a person who keeps looking for his lost keys under the streetlight because it is too dark at the place where he might have dropped them.

Understandable as it may be, this negative attitude of science is inexcusable. After all, all scientific evidence is based on human experiences; the human mind is the ultimate measuring apparatus. Yet the nature of the Subject of Cognizance is never raised as a scientific issue! This is like using a telescope to investigate the heavens and never bothering to inquire what it is.

Present-day science is based not only on the principle of objectivation but on human experiences as well. Since experiences are subjective and, as experiences, cannot be objectivized, the very basis of the scientific endeavor is excluded from the domain of scientific inquiry. And since self is, by definition, unobjectivizable, the question "Who am I?" is the least accessible to the scientific methods.

This need not be a limitation of science per se. It may well merely be a characteristic of science as it is currently practiced. It is possible that *the integration of the objective and subjective domains within the context of the scientific endeavor will be the next decisive step in the evolution of science.*

This idea is not new. The possibility of such integration goes back to the Greeks, who were interested in both the subjective and the objective domains and saw both as proper fields of study. In our own time it was Schrödinger who saw the need not only to include the question "Who am I?" within the scientific enterprise but to give it the place of honor:

> I consider science an integrating part of our endeavor to answer the one great philosophical question which embraces all others, the one that Plotinus expressed by his brief: τίνες δὲ ἡμεῖς—*who are we?* And more than that: I consider this not only one of the tasks, but the task, of science, the only one that really counts [italics in original][2]

Does Schrödinger's statement call for a debunking of science as we know it? Does it call for a cessation of the exploration of the phenomena of nature in favor of the penetration, in some way, into the subjective domain? If what we are really after is an exploration of Plotinus's question, what is the point of doing what we do in science, i.e., explore the sensible world around us, from galaxies and stars to viruses and molecules?

It seems to me that, with a proper shift of perspective, present-day science can, indeed must, be included in the inquiry into our own nature. To see how, let us digress for a moment and examine our relationship to our dreams.

If I wish to explore my inner world, should I investigate my dreams? Well, it depends. If I am under the illusion that my dreams are not dreams but real happenings in the outer world, their investigation will not tell me anything about myself. Such an investigation would involve a domain of reality that may be fascinating to explore but has nothing to do with *me*. If, however, I understand, as we all do, that dreams are projections of my own mind, their investigation can be invaluable in the exploration of my inner world—a fact that is, of course, the cornerstone of many methods of psychotherapy, from Freud's psychoanalysis to Perls's Gestalt Therapy.

I venture to guess that if Plotinus were alive today he would have been very interested in the discoveries of our science *as revelatory of the inner world*. For Plotinus, you recall, the sensible realm had no independent existence; it was an expression and a reflection of the Soul.

2. Self and Ego

Well, then, who is the Subject of Cognizance? Let us begin by examining ordinary perceptions.

I look at a tree. My attention is focused on it, and not on myself, the one who sees the tree. In the context of an act of perception, my knowledge of myself is indirect, inferential. I *know* that I see an object; hence I must be present, seeing that object. From another point of view, however, the knowledge that *I am* is so elemental that it needs no proof. I am not included in the field of my attention, which is focused on the tree, because *I am the source of this attention*. The impossibility of perceiving oneself as an object in the subject/object mode is analogous to the impossibility of seeing light when we look at objects. When we see an object, we see it because light from the object is reaching our eyes. The impingement of light on our retinas enables us to see; but what we see—*by the light*—is the object, and not the light.

In any act of perception there is a demarcation line between the subject, the one who sees and is not seen, and the object, which is seen. This demarcation line shifts easily as one moves from one act of perception to the next. In particular, if I am interested in knowing the subject, I can include it, i.e., include "the one who sees the tree," in the field of my attention. I am looking now at "myself looking at a tree." By this very process, however, this self has become an object! But since as long as I perceive in the subject/object mode "myself" is precisely the subject, the self that I now see as an object is certainly not the Subject of Cognizance.

This state of affairs is what the protagonist in Niels Bohr's favorite story, Poul Martin Møller's "The Adventures of a Danish Student," finds so confusing. The protagonist is referring to thoughts rather than perceptions, but the problem is essentially the same. Here is Bohr's favorite passage:

I get to think about my own thoughts of the situation in which I find myself. I even think that I think of it, and divide myself into an infinite regressive sequence of "I's" who consider each other. I do not know at which "I" to stop as the actual one, and in the moment I stop at one, there is indeed again an "I" that stops at it. I become confused and feel a dizziness, as if I were looking down into a bottomless abyss.[3]

The end point of the passage is, however, the beginning of the real inquiry. The contemporary philosopher Anthony Damiani puts it this way:

When you look, when you introspect into yourself, don't you see that there is a nothingness in you? Or do you keep skirting around it? Just look into yourself, just stop for a moment. Haven't you noticed that it's always there, the emptiness that you are running away from?[4]

Damiani proceeds to point out that this nothingness, this empty center, is none other than one's real self; he uses the dream analogy to explain:

Think of a dream. We have a series of thoughts one after the other, but they all revolve around the unknown center—the person who's having the dream. The entity in the dream has the feeling that he or she has or is real being ... [but] the reality is coming from the person. The congeries of thought is what the dream is all about. And they revolve around the reality of the person. But if you look at the congeries of thought and you expect to find a real ego there, all you'll find is one thought after the other. In other words, there's a level of higher being which we're equating with the person, and there's a lesser level or lower level of being and we are saying that's the dream. The person is immanent in the dream. That immanence gives the ego the feeling that it is *some* **thing**. That's a misconstruction, a misunderstanding. . . . It is only the immanence in the dream of the person having the dream that . . . makes it possible for a dream ego to arise. The center of the vortex is *not* neutral. The center of the vortex is the presence of the "I AM" in the matrix of possibilities [italics and bold in original].[5]

The dreamer, that is, the real person who is in bed dreaming, can never be one of the figures in the dream. And yet the dream figure he is identified with seems to have a feeling of the person's "I" immanent in it. This feeling is projected onto it by the dreamer. In the same way, self, or soul, will never be found as an object that the mind can behold. I have a sense of my ego as a self-subsisting entity not because it is a self-subsisting entity but because it is projected onto my ordinary mind by my real self.

3. The Oneness of Mind

In Chapter 9 we mentioned Schrödinger's 1956 Tarner Lectures, which appeared as a book entitled *Mind and Matter*, and discussed the third lecture, "The Principle of Objectivation." The fourth lecture, just like the third, is a gem. Entitled "The Arithmetical Paradox. The Oneness of Mind," it deals with the following issue: Ordinarily we live under the impression that there is one objective, real world and many "selves." Each one of us seems to have his or her own conscious mind, his or her own "self." Is this impression valid? How many of us are there?

The two aspects of this impression, the apparent existence of one objective, real world and the apparent presence of many selves, are interrelated. The objective world does not exist on its own; it is, rather, the result of the process of objectivation, a process that is based on a bifurcation of the real world into an objective world and a Subject of Cognizance. If, however, there are many minds, i.e., many Subjects of Cognizance, why do they all seem to share one objective world?

Schrödinger responds to the challenge of this question in one bold stroke: "There is obviously only one alternative, namely the unification of minds or consciousnesses. Their multiplicity is only apparent, in truth there is only one mind."[6] This strange assertion seems to run against the clear evidence of experience. Isn't it obvious that there are many of us?

No. It is obvious that there are many bodies, as well as many egos, but living, as we do, in Plato's cave, we mistake shadows for realities. We confuse ego with self; we translate our evidence for the existence of many egos into a conclusion about the existence of many selves. Schrödinger's contention is that just as there is only one dreamer behind the many dream figures, there is only one mind, one self, behind the many egos. Schrödinger proceeds to support this claim with logical analysis as well as with the evidence of direct experiences.

Logical analysis: First and foremost, the statement that there is only one mind does resolve "the arithmetical paradox"—the apparent existence of one objective world and many minds. There is, however, another compelling piece of evidence:

The doctrine of identity can claim that it is clinched by the empirical fact that consciousness is never experienced in the plural, only in the singular. Not only has none of us ever experienced more than one consciousness, but there is also no trace of circumstantial evidence of this ever happening anywhere in the world . . . we are not able even to imagine a plurality of consciousnesses in one mind. We can pronounce these words all right, but they are not the description of any thinkable experience. Even in the pathological cases of a "split personality" the two persons alternate, they never hold the field jointly.[7]

This argument is followed by biological and psychological data that drive the point home.

Schrödinger does not claim the discovery of the doctrine of the oneness of mind for himself. Far from it. He quotes both ancient and modern sages, invoking the wisdom of East and West. "My purpose in this discussion," he declares, "is to contribute perhaps to clearing the way for a future assimilation of the doctrine of identity with our own scientific world view, without having to pay for it by a loss of soberness and logical precision."[8]

The doctrine of the oneness of mind accords well with Plotinus's view on the nature of matter. For Plotinus matter is like a mirror in which the noumenal world is reflected (see Chapter 17, Section 4). Each one of us acts like such a mirror; each brain produces its own reflection. These are, however, many reflections of one Mind; and, needless to say, the true reality is the Mind and not the reflections.

Direct experiences: As I have just mentioned, the doctrine of the oneness of mind has been enunciated by contemplatives and thinkers throughout the ages, and throughout different cultures. Schrödinger refers us to Aldous Huxley's book *The Perennial Philosophy* for numerous examples. Here, however, I would like to invoke a rather unlikely witness to the truth of the doctrine.

Freeman Dyson belongs to the second generation of quantum physicists, the generation that was active after the Second World War. One of the issues that was left unresolved by the founding fathers of quantum mechanics was the development of a full-fledged theory of quantum electrodynamics—a formalism for the calculation of the effects of electromagnetic interactions in the atomic and subatomic domains. The problem was solved, in the late forties, by four physicists of this second generation, Julian Schwinger, Richard Feynman, Sinitiro Tomonaga, and Freeman Dyson. In the remainder of his career Dyson was involved mostly with imaginative applications of the basic theories of physics, rather than with the development of the theories themselves. He focused his attention on the use of science for the exploration of new possibilities for humanity, such as space colonization.

Basic philosophical issues hardly feature in Dyson's writings; in particular, I am not aware that he ever mentioned, in his writings, the idea of the oneness of mind. And yet . . . Kenneth Brower, who wrote a book about Freeman Dyson and his son, George, a master canoe builder, mentions the following episode:

> Freeman told us that when he . . . was fourteen he had started a religion. Unhappy with the Christian notion that the heathen are doomed for reasons out of their control, he had begun a sect of his own. "I was convinced suddenly that all people are the same. We are all one soul in different disguises. I called it Cosmic Unity. . . . I seem to remember that I even had a convert. Cosmic Unity lasted about a year, I think."[9]

I have never met Freeman Dyson and have never talked to him about this episode. For all I know, he may dismiss it as an adolescent fancy. For me, however, the word "suddenly" in the quotation above expresses an intuitive leap of the intelligence into the noumenal dimension. The sentence "We are all one soul in different disguises" is a precise enunciation of the idea of the oneness of mind. This gifted fourteen-year-old boy suddenly tapped into the Universal Mind and emerged with the same truth that was expressed in different forms by the sages and thinkers whom Aldous Huxley quotes in *The Perennial Philosophy*.

4. Transcending the Subject/Object Mode

We ended Chapter 9 by quoting an enigmatic statement of Schrödinger's, the closing statement of his objectivation lecture:

> The world is given to me only once, not one existing and one perceived. Subject and object are only one. The barrier between them cannot be said to have broken down as a result of recent experience in the physical sciences, for this barrier does not exist.[10]

Now we are in a position to comprehend, or at least glimpse, the meaning of the statement.

As you may recall from Chapter 17, Section 4, Plotinus's world-view involves a hierarchy of levels of reality, the highest of which he calls "the One" or "the Supreme." The One is followed successively by Nous, the World Soul, individual souls, nature, and the sensible world. This sequence is not a sequence of entities but is, rather, a sequence of levels of possible experiences of the one reality. As one contemplates, penetrating deeper and deeper into the nature of reality, one leaves the sensible world behind, transcends the subject/object mode of perception, and experiences one's soul. As one's state of contemplation deepens, as one becomes more and more "self-encentered," one comes to experience the nature of one's soul as being one with the World Soul, which is, in turn, an aspect of the Nous. And, ultimately, the Nous is not other than the One.

This last paragraph was glibly written. The words come easily, but, as a rule, the experiences they indicate involve what Plato calls in his cave allegory "a steep and rugged ascent." When experiences of the noumenal do come, however (and there are different levels and different kinds of such experiences), they come with a sense of utter certainty and with the feeling that they are impossible to communicate. Heisenberg's experience in Prunn Castle is an apt example. It came to him with utter certainty, but all he could tell us was that he found his link with "the central order."

Experiences of the noumenal world are indescribable, yet many of those who have had such experiences have tried to convey them as best they could. In a sense, the whole of Plotinus's system of thought is just such an attempt. As Anthony Damiani puts it:

> The Enneads are formulations of Plotinus's intuitive realizations of ultimate truths to which only our inner being can respond. We must let the logos in our soul absorb the impact and assimilate the meaning of his intuitions before allowing our critical and egotistical intellect to pounce upon them. In other words, our mental activity must be stilled so that the silence within can receive the passage in question without coloration.[11]

The higher the level of the experience, the more challenging the task of expressing it. In Plotinus's case, the challenge culminates in his utterances that relate to the One and to our relationship to it.

In our self-seeing There [i.e., in the Supreme], the self is seen as belonging to that order, or rather we are merged into that self in us which has the quality of that order. It is a knowing of the self restored to its purity. No doubt we should not speak of seeing; but we cannot help talking in dualities, seen and seer, instead of, boldly, the achievement of unity. In the seeing, we neither hold an object nor trace distinction; there is no two. The man is changed, no longer himself nor self-belonging; he is merged with the Supreme, sunken into it, one with it: centre coincides with centre, for centres of circles, even here below, are one when they unite, and two when they separate. This is why the vision baffles telling; we cannot detach the Supreme to state it; if we have seen something thus detached we have failed of the Supreme which is to be known only as one with ourselves.[12]

The challenge of coming to the ineffable knowledge of who I really am is the same as the challenge of coming to the ineffable knowledge of the One. This is so because, ultimately, *I am the One*. If one accepts that "the One" is a name indicating the real, nameless source of what is, rather than an abstract concept, then, oddly enough, the statement "I am the One" can be proved: If I were not the One, then the level of "the One" would have consisted of at least two items, me and the One, rather than there being truly one.

Toward the end of the fourth Tarner lecture, "The Arithmetical Paradox. The Oneness of Mind," Schrödinger comes back to the theme of the double role of mind. On the one hand, mind is the Universal Mind; on the other, it is "my mind" or "your mind," a mind associated with a finite being, a being who was born and who will die. He invokes two metaphors to explain this double role: first, Homer, the author of *The Odyssey*, as compared with the bard, a character in *The Odyssey*, who is believed to represent Homer; and, second, a drawing by Dürer, which includes, among the many characters, a depiction of the artist.

Such metaphors help us understand the double role of mind in relation to the sensible universe. On the one hand, Mind is the stage that is prior to the display of phenomena. If we adopt the terminology of current cosmology, we can say that it existed before the Big Bang and will continue to exist after the Big Crunch, if there will be one. On the other hand, this display is reflected, however imperfectly, in my mind and yours. Yet, just as the ultimate reality of a dream character is the mind of the dreamer, our individual minds are, ultimately, the one Mind, the One. For the individual, therefore, the quest for the experience of Oneness is the quest for the experience of his or her deepest identity, as well as the experience of the nature of reality.

5. The Limitation of Science

The *scientific* quest for the understanding of the nature of reality is beset by the following difficulty: Very deep truths can be known. Very high levels of reality can be experienced and comprehended, and the experiences come with a sense of utter certainty. Since they transcend duality, however, they cannot be conveyed in the language of ordinary human intercourse, a language that is based on contrasts and distinctions. As a consequence, such experiences are useless, even meaningless, from the point of view of present-day science, the practice of which is based on communication in plain language.

The issue of the relationship between science and language occupied Niels Bohr throughout his life. He said:

As the goal of science is to augment and order our experience, every analysis of the conditions of human knowledge must rest on consideration of the character and scope of our means of communication. The basis is here, of course, the language developed for our orientation in the surroundings and for the organization of human communities.[13]

Bohr knew very well, however, that Truth does not yield easily to this mode of communication. His favorite quotation was a pair of lines from a poem by Schiller:

Only wholeness leads to clarity
And Truth lies in the Abyss.[14]

The framework of complementarity was Bohr's solution of the problem. He accepted the fact that what can be known about a quantum entity, such as an electron or an atom, cannot be contained in a single description. At least two seemingly contradictory descriptions are needed. The two descriptions are really complementary, rather than contradictory, because they apply to different circumstances. Each expresses an aspect of the truth; neither conveys the whole truth. The truth lies in the abyss between them and can be comprehended experientially to the extent that one can hold in one's mind, simultaneously, both descriptions, as well as the different circumstances in which they apply.

The framework of complementarity is paradoxical in the following sense: Discovered as a result of an attempt to describe quantum systems in plain language, it is a precise expression of the impossibility of such a description.

The discoveries of impossibilities have often been important milestones in the history of physics. The understanding of the concept of energy was clinched when it was discovered that it is impossible to create or destroy energy, that any apparent creation or destruction is really a transformation of energy from one form to another. Heisenberg's uncertainty principle is a statement of the impossibility of measuring simultaneously both position and momentum to an arbitrarily high level of accuracy. The framework of complementarity is a discovery of the impossibility of applying a single description to a quantum system.

The framework of complementarity can be considered, then, a first step in the direction of transforming science into a mode of exploration that can incorporate ineffable truths. However, the difficulty of applying the scientific method to the exploration of truths that are revealed in very deep experiences goes much deeper. Two issues are involved: first, the data, i.e., the raw material of the scientific inquiry, and, second, the processing of the data.

The data: Newtonian physics is based entirely on sensory evidence. Its spectacular success led scientists to consider other types of experiences, such as feelings or intuitions, irrelevant. The leading philosophers of the Enlightenment, Locke, Hume, and Kant, were so awed by the success of Newtonian physics that they too denied the validity of all but sensory experiences as revelatory of truth. As Locke put it, "There is nothing in the intellect that was not previously in the senses."[15] Kant challenged Locke by asserting that the mind does have *a priori* structures for the assimilation of sensory experiences, but he agreed with Locke, Hume, and the other empiricists that sensory perception is the only conceivable source of certainty in regard to knowledge of the universe. Richard Tarnas summarized the situation succinctly:

> Where Plato had held sensory impressions to be faint copies of Ideas, Hume held ideas to be faint copies of sensory impressions. In the long evolution of the Western mind from the ancient idealist to the modern empiricist, the basis of reality has been entirely reversed: Sensory experience, not ideal apprehension, was the standard of truth—and that truth was problematic.[16]

Kant's analysis led him to the conclusion that the phenomenal world is a creation of the human mind and that, moreover, only the phenomenal world is knowable. The only knowledge we can have, then, is knowledge of our own minds. We cannot know reality as it is in itself.

Kant's analysis is valid *if sensory experiences are the only valid source of knowledge.* But what if Plato and Plotinus are right? What if deep truths become available precisely when one finds a way to leave sensory experiences behind and transcend the subject/object mode of perception? What if present-day science ignores such experiences not because they are invalid but because their ineffability makes it impossible for present-day science to incorporate them within its ken?

According to William James, "we feel neither curiosity nor wonder concerning things so far beyond us that we have no concepts to refer them to or standards by which to measure them." As an example he mentions an incident that occurred when Charles Darwin visited Tierra del Fuego. "The Fuegians, in Darwin's voyage, wondered at the small boats, but took the big ship as 'a matter of course.'"[17]

The processing of the data: The scientific methods of processing the data include induction and deduction. These are two varieties of *discursive reasoning* and are operations of the logical mind. To be sure, the process of scientific discovery involves intuitive leaps that transcend logic, but such experiences are events in the scientists' inner world and, as such, are not part of science per se.

Is discursive reasoning the only valid mode of processing the evidence? From the contemplative point of view, just as sensory experience in the subject/object mode is a relatively superficial way of receiving impressions, discursive reasoning is a relatively superficial way of seeking knowledge. Plotinus distinguishes between a superior, contemplative mode of knowing, knowledge by the vision of unity with the Supreme, and an inferior mode, knowledge by reasoning. In the first mode the knowledge and the experience are one. The second involves a separation between the data and its processing. In the first mode "the soul looks upon the wellspring of Life, wellspring also of Intellect, beginning of Being, fount of Good, root of Soul."[18] One phase of the Soul stays connected with this vision always, while another phase of the soul tends to lose it.

> It is the other phase of the soul that suffers, and that only when we withdraw from vision and take to knowing by proof, by evidence, by the reasoning process of the mental habit. Such logic is not to be confounded with that act of ours in the vision; it is not our reason that has seen; it is something greater than reason, reason's Prior, as far above reason as the very object of that thought must be.[19]

To be sure, neither Plato nor Plotinus is opposed to discursive reasoning; their own writings are replete with logical arguments. They merely wish to establish the relative value of discursive reasoning in the quest for knowledge. It has its place, but this place is nowhere near the place of the beatific vision, the experience of merging with the One.

But what about us, poor devils, who have never experienced the vision? What are we to do? At the very least, says Plotinus, we must admit the existence and superiority of the noumenal mode of being and knowing, since *the order of nature is evidence of a source of order.*

> Even those who have never found entry [to the noumenal world] must admit the existence of that invisible; they will know their source and Principle since by principle they see principle and are linked with it, by like they have contact with like and so they grasp all of the divine that lies within the scope of mind. Until the seeing comes they are still craving something, that which only the vision can give; this Term, attained only by those who have overpassed all, is the All-transcending.[20]

Present-day scientists and philosophers seem to be, in relation to the Truth, as the Fuegians were in relation to Darwin's ship. The Truth is here; it has been experienced repeatedly, with a sense of utter certainty. But science, based, as it is, on sensory experience and discursive reasoning only, cannot deal with it. The resolution of this impasse may well turn out to be the scientific challenge of the twenty-first century.

6. The Quest of Physics: Grand Unification

The removal of the Subject of Cognizance, that is, ourselves, from "the domain of nature that we endeavor to understand" is the removal of life from nature. Hence the possibility of a direct experience of the vision of the One, the source of life, is absent from the current scientific paradigm. But the yearning for this vision is present within any soul that has not had the experience. This yearning may be unconscious, but it is present. Since its pursuit within science is excluded by the principle of objectivation, it is manifested there in the only form science can give it, namely, as a search for grand unification, a single theory that explains all phenomena.

The search for such a theory is not new. When Newton published his Principia, three hundred years ago, it was hailed as the grand unification. No less a mind than Kant's was certain that the Newtonian framework was the ultimate truth concerning the phenomenal world. This stance was confirmed repeatedly by the successes of the Newtonian system throughout the eighteenth and nineteenth centuries. The series of successes reached its culmination in the latter half of the nineteenth century, when James Clerk Maxwell formulated his theory of electromagnetism. Maxwell's theory seemed to fit very well into the Newtonian framework, provided one was willing to accept the idea that space is absolute and filled with an unknown substance called "ether."

As the nineteenth century was drawing to a close, the physicist A. A. Michelson affirmed the vision of the Newtonian framework as the final theory in no uncertain terms. He claimed (in 1894) that there were no more fundamental discoveries left to be made. By then two minor problems clouded the vision just a bit. One was, paradoxically, the result of Michelson's work: the Michelson-Morley experiment of 1886, which showed that the speed of light was unaffected by the movement of its source relative to the ether, an experimental fact that was hard to account for. Second, calculations of the amount of radiation emitted from hot, glowing objects yielded infinity as an answer—an obvious absurdity. The consensus was, however, that these two disturbing issues would soon find their solutions within the Newtonian framework. This hope proved vain: The Michelson-Morley result was finally explained in 1905 by Einstein in his Special Theory of Relativity, a theory that did away with the Newtonian paradigm regarding space and time, and the radiation problem was solved in 1900 by Max Planck, who introduced "the quantum hypothesis," which started the quantum revolution, the revolution that did away with the Newtonian conception of matter. By the 1920s the Newtonian paradigm was in shambles.

Since then much has happened in the world of physics. Two new types of interaction have been discovered: the "nuclear interaction," which is dominant inside nuclei of atoms, and the so-called "weak interaction," which features in radioactive processes. A theory that unified the electromagnetic and weak interactions was proposed in 1979 and verified in the 1980s. A search is now underway for a theory that will include the nuclear interaction as well and, one hopes, gravity. Such a theory will fulfill the hope for "grand unification."

What are the chances for such grand unification? The most promising current candidate, string theory, is so beset by mathematical complexity and conceptual ambiguities that it is nowhere near the possibility of experimental testing. Yet some people, Stephen Hawking included, are cautiously optimistic. To be sure, the historical lesson of the Newtonian framework has not been lost. Even the optimists warn us that, should a viable grand unification scheme be found, it will always be a potential candidate for extinction due to the emergence of new data.

A theory incorporating the ideal of grand unification may or may not be in the offing. It seems to me impossible, however, that even if such a theory were found, it would provide a final framework for the understanding of reality. Any scientific theory is a conceptual framework, based on the principle of objec'ivation. Failing to incorporate the Subject of Cognizance, it represents a dead image of a living universe. Even if such an image were consistent, for a while, with the scientific data, the alive aspects of the universe would be bound to show up sooner or later and prove it wanting. Moreover, as Hawking pointed out, such a theory would not provide the answer we are really seeking. A theory is "just a set of rules and equations," he writes, but "what is it that breathes fire into the equations and makes a universe for them to describe? The usual approach of science to constructing a mathematical model cannot answer the questions of why there should be a universe for the model to describe."[21]

I believe that when future historians of Western civilization look back on the last five hundred years, they will look at our science from the perspective of a unified view of the noumenal and the phenomenal realms within the framework of a new science and see our exclusive reliance on sensory experience and discursive reasoning, and our adoption of the principle of objectivation, whereby a living universe is treated as a vast collection of chunks of dead matter, as a transitional aberration. Such a transition may be necessary, even inevitable; but we may end up saying about it what Einstein said to Heisenberg about his (Einstein's) adherence to Mach's philosophy, an adherence that was pivotal to the process that led to the discovery of Special Relativity: "Possibly I did use this kind of reasoning, but it is nonsense all the same."[22]

EPILOGUE

There is a special mood one sometimes gets into when a long and meaningful project is drawing to a close. Sitting at my desk, at my home in rural Vermont, facing the computer screen where the word "epilogue" appeared in bold letters, I felt a mixture of satisfaction, sadness, some regret, the source of which eluded me, and some laziness. Through the window a lush, sunny scene beckoned: trees, flowers, clouds, a light blue sky. The moment was alive, just as it was; meaning was not an issue. Yet, as soon as I brought my attention back to thinking, and to issues relating to the conclusion of this book, I re-situated myself in this time, in this civilization, in this culture, and the well-known existential questions came right back.

As long as I share the world-view of this postmodern era, I see no guidance for the mind, no signpost, no beacons. Nature is impersonal, devoid of life, meaning, and value. In the domain of human intercourse, division and confusion reign supreme. In response to this challenge, a thousand prophets proclaimed a thousand little truths; I gulped them all down and stayed hungry. Does the material of this book provide me with a way out?

I felt unsure, and the more I explored these existential issues with my mind, the more unsure I became. And yet, as my attention wandered away from thought, as I watched a small branch moving gracefully in the breeze, meaning returned, ineffable to be sure, but meaning nonetheless.

Turning my attention from the view out the window to the papers on my desk, I started rummaging through a pile of old quotes I copied years ago and came across a crumpled piece of paper, on which I could discern the words "Looking beyond all objects, both inner and outer, and into their source . . ." Just then I heard a car moving into the driveway and recognized the vivacious voices of Julie and Peter. I stuck the piece of paper into my pocket and went out to greet them. We settled down in the shade under a tree and took in the view.

After a while, Peter said: "I'm glad you wrote this book. It deals with an important and difficult issue: the fact that our world-view is made up of contradictory bits and pieces. As I read your book, I didn't feel that it resolved this difficulty, but it did provide a direction. It's like . . . how shall I put it? My hunger hasn't been satisfied, but I know where the food is; I know where to look for it."

"I want to hear more about this direction you speak about," Julie responded. "You were talking about the direction of a solution, the direction of a world-view within which we can feel rightly oriented. But, before we get to the solution, it's probably a good idea to talk about the problem.

"Remember Dante's *Divine Comedy*? Seven hundred years ago Dante's vision provided the Western mind with a comprehensive world-view. We humans had a central place in the universe; we knew it, and we knew what this place was. Since then three major intellectual revolutions, connected with the names of Copernicus, Descartes, and Kant, have deprived us of any kind of significance and certainty. Richard Tarnas delineates it well in his book *The Passion of the Western Mind:* With Copernicus we lost our place at the center of the physical universe. With Descartes we lost our place in nature altogether: We became disembodied minds in a lifeless universe. Moreover, Descartes, with his radical doubt, started a movement that led eventually, with Kant, to the loss of the conviction that we can know anything about that which really exists, anything that is not the creation of our own minds. I think that before we even try to talk about the solution, the emergent world-view, we should try to deal with these very losses."

"Well," I said, "the real question is this: How significant are the findings of Copernicus, Descartes, and Kant? Strange as it may seem, we can approach this question by taking some hints from quantum mechanics. To be sure, the quantum theory can give us only hints. Quantum mechanics itself is not a world-view, and, since it was born out of the classical scientific paradigm, it participates in this paradigm to a large extent, but it does transcend this paradigm at some essential points, and it does provide hints of the way out. Heisenberg pointed out, for example, that the concept of matter as it emerges out of quantum mechanics is really a Platonic one. Following such leads, I took further hints from Whitehead, whose system of thought is essentially Platonic and can incorporate the findings of the quantum theory very well. And I found Schrödinger and Plotinus very helpful too, as you know.

"Now, Plato and Plotinus answered the Copernican challenge many centuries before it was offered. For Plato the sun, and not the earth, is the central entity in the physical universe; and this emphasis on the sun, rather than on the earth, does not diminish our cosmological role at all. Our role has nothing to do with the place of our bodies in the physical universe. Our role hinges on our relationship to the World Soul, to the Nous. It is dependent solely on the depth of our experiences."

"I see what you're getting at," Peter added. "I hear you saying that the limitations that Descartes and Kant are said to have discovered are also based on invalid assumptions. These philosophers assumed, first, that discursive reasoning, which uses the dualistic, logical mind, is the only vehicle, and the right one, for coming to certain knowledge about the universe, and, second, that sensory information is the only relevant data."

"That's right," I responded. "From a Platonic point of view, the validity of which is fortified by records of experiences of major figures in our civilization, including some of the greatest scientists, these assumptions are plainly wrong."

"When I think about our current Western civilization," Julie interjected, "what impresses me most is the tremendous amount of suffering inherent in it, like feelings of meaninglessness, alienation from nature and the transcendent realm, boredom, and so on."

"Well," I responded, "this suffering too is due to the unnecessary limitation that the Western mind imposed on itself by putting its trust in the sensory input and in the analytical approach, which is dualistic through and through. Whether we are aware of it or not, our deepest need is to be in relationship with the One, the ultimate unifying principle. But the One transcends all distinctions; it cannot be related to in the dualistic mode. The human mind does have the possibility, however, of connecting with the One through a higher, numinous mode of being, a mode of being that begins with a still mind. As St. Augustine puts it, 'if only their minds could be seized and held steady, they would be still for a while, and, for that short moment, they would glimpse the splendor of eternity which is forever still.'[1] Our alienation, then, is ultimately an alienation from our own possibilities."

"Yes," Peter responded, "I see that, but I still feel stuck. I see that the so-called proofs that it is impossible to know the essential truths with certainty are faulty, but *I don't know the essential truths with certainty.* This is what I meant when I said that my hunger is still present."

At that moment we were distracted by the appearance of a deer. It moved, unhurriedly, out of the woods and toward the vegetable garden, a hundred feet away from us. We stayed very quiet. Nevertheless it must have felt our presence. It turned its head toward us, gave us a curious look, and bounded gracefully back into the woods.

"Wow!" Peter said. "Amazing!"

I looked at Julie and saw a mischievous look in her eyes. I knew we were about to have some fun. "Just now, when you looked at the deer," she said, "*did you have any doubt?*"

Peter was startled. "Doubt about what? What are you talking about?"

"Why, about the universe, about knowledge, about valid and invalid assumptions . . ."

"Of course not, I wasn't thinking about that!"

"There you are!" Julie concluded, as if she had made her point and there was nothing more to be said.

Had this exchange taken place at the beginning of the book, Peter would have been puzzled, disturbed, and even angry. He would have picked a fight with Julie, and it would have taken me quite a few paragraphs to quiet them both down. By now, however, all three of us knew each other rather well, and Peter respected Julie's way of trying to tell him things. He turned his head, first toward Julie and then toward me, and, getting no response, he closed his eyes and sat quietly. I could not tell whether he was pondering or contemplating. Finally he opened his eyes and smiled. Julie beamed. "Tell us all about it," she said.

"Well," he responded, "at first I tried to figure out what you had just said. Slowly, however, I became quieter and just lost interest. I noticed that, while none of my doubts got resolved, *the doubts themselves ceased to be all that important.* Then I understood that the whole existential quandary, alienation, boredom, meaninglessness, all the way to the 'death of God,' all of it is due to a wrong use of the mind."

"What do you mean?" I interjected.

"You of all people should know what I mean," he retorted, a bit testily, I thought. "Although others have said it before you, *I* got the idea from *your* book. We are accustomed to use the dualistic part of the mind for everything. But this part of the mind is not appropriate for the comprehension of ultimate questions. It is simply the wrong tool. If you try to carve a delicate little piece of sculpture with a kitchen knife, you are bound to be frustrated. And if you are used to blaming others rather than looking into yourself, examining yourself, as Socrates recommended, you will end up concluding that God is against you, or that the universe is meaningless and so on. Eventually you will lose interest even in that and just be bored."

"We understand that. Now go on," Julie spoke softly.

"As I became more and more quiet," Peter continued, "it dawned on me that the only place to look for both Meaning and Experience is within myself, the Subject of Cognizance, away from all objects, inner or outer. Then something happened. It was real and powerful, but I can't describe it."

"I've just come across a piece of writing," I said, "which seems to convey the flavor of your experience. I copied it long ago, and I can't remember now who wrote it." I pulled the piece of paper out of my pocket and read aloud:

Looking beyond all objects, both inner and outer, and into their source, beyond trees and mountains, beyond thoughts and feelings, or, rather, looking at inner or outer objects with such acute attention that the subject/object mode melts away, another reality opens up. Just as the physical world around us has been revealed to be of unimaginable dimensions, we enter now into the inner world, which is of unimaginable depth. The individual soul, the World Soul, the Nous, the One—these are but names; the range and depth of the potential experiences they indicate, however, leads beyond the individual subject, into the cosmological dimension, and beyond it too.

Ineffable and numinous, this reality is the source of meaning and experience in the so-called ordinary reality. In its light, the postmodernist claims of relativism as a fundamental truth (a stark contradiction, this!) are dispelled, without argument, as darkness is dispelled by light.

The Experience feels dichotomous. The other reality is really other than the familiar one. But this feeling of dichotomy comes *from this side*; it arises when the Experience is over, and the ordinary mind evaluates it. *From its own side*, the reality that opens up is beyond dichotomies; it feels like home— and yet we cannot stay there.

The fact that we cannot stay there is indicative of the real challenge—the bringing together of the two, the infusion of the phenomenal with the noumenal, and the expression of the noumenal in the phenomenal. This means, on the one hand, *seeing* the inner dimension through the phenomena that present themselves, seeing the ineffable through the words, through the formulae, through the symbols, and on the other hand, *creating* phenomena through which the noumenal can shine, expressing the ineffable in words and symbols.

The intensity of such an engagement makes existential doubts irrelevant; and then there comes a time when one knows, without a doubt, that this place, between the phenomenal and the noumenal, is one's place in the universe.

"Yes," Peter said slowly, "that's more than I can say, but it is in line with my experience, and it does make sense."

"I feel the same way," Julie added. "And another aspect of the issue stands out for me. The noumenal world is the *source* of meaning. The source is ineffable, while meaning can be expressed in ordinary language, up to a point." She stopped—she seemed unsure of something—then resumed her explanation. "But when I try to formulate the new meaning, the direction of this new emerging paradigm, I come to a somewhat paradoxical point of view. The ideas of an alive, multileveled universe in which we human beings participate organically provide me with a valid direction, an intellectual orientation that makes sense, but we cannot discard the lessons of either postmodern relativism or Bohr's complementarity. I feel that the relativists are right in the sense that any paradigm expressed as a conceptual system is relative to the culture that formulated it. It follows that any conceptual system is limited in its capacity to be a carrier of the essential truths, the final answers about the universe and about ourselves. So we need to look at any paradigm, including the one that is utterly convincing to us, as just one aspect of the understanding of the truth. It has to be complemented by something else."

"Which is . . . ?" It seemed as if Peter were holding his breath.

"Why, the ineffable experience, the source of the paradigm! There is no substitute for the Experience itself. The hunger you were speaking about can only be satisfied by the Experience, and yet, once the Experience takes place, there is a real need to express it as a paradigm. The essential function of the paradigm is to direct the mind toward the possibility of the Experience. So, in a strange sort of way, the ultimate role of a paradigm is to negate itself."

Peter became thoughtful. "That idea makes me uneasy," he finally said. "Not alienated or hopeless, certainly not bored—just uneasy."

"It should." Julie was relentless. "There is a hidden message in what I just said: The Western mind in the second part of the second millennium has been extremely stupid."

That was enough to startle both Peter and me. "What do you mean?" we said, almost in unison.

"Why, the great challenge of exploring what the universe really is, its numinosity, has, is, and will always be present, but the Western mind resolutely turned away from this challenge, proclaimed shadows as realities, and expected to find the One among the shadows. This is stupid. Look at Ernst Mach. This superficial philosopher exerted such an influence at the end of the nineteenth century and the beginning of the twentieth that both young Einstein and young Heisenberg fell under his spell!"

"And yet," Peter responded, smiling, "Mach's influence was crucial for the creation of both Einstein's theory of Special Relativity and Heisenberg's quantum mechanics. How do you account for that?"

Julie stopped in her tracks. "I don't know," she said. "Let's look into it."

"I appreciate your question," I said to Peter. "How indeed is it possible that a philosophy as patently wrong as Ernst Mach's would exert such a strong influence on brilliant minds and that, moreover, *this influence would be beneficial to the progress of science!?*"

We sat quietly, pondering the question. A few minutes went by, and then Julie spoke. "Perhaps we will solve the riddle," she said, "if we look at Peter's question along the lines of the new, emergent paradigm and regard the evolution of the Western mind as the development of an organism. When my little brother was nine or ten, he went through a period of being really obnoxious. At the time I couldn't stand it, but in retrospect I see it as a developmental phase he just had to go through. Similarly, it is possible that as science, art, or literature develops, it may need to function on the basis of wrong ideas for a while."

"Yes, I can see that," Peter responded. "Actually, this explains why so often throughout history the most influential thinkers were not necessarily the most profound. If a civilization is an organism that has needs, then the thinkers who would occupy center stage would be those who make conscious the ideas it needs at the time. Other thinkers, who are ahead of their time, would be ignored."

"Yes," I said, "Ernst Mach is a case in point. He brought to science an input it really needed. Mach did his work on the background of the speculative German philosophy of the late nineteenth century, most of which consisted of meaningless verbosity. Mach's beneficent influence consisted primarily in his insistence that we know precisely what we are speaking about. Einstein accepted this demand and applied it to the concept of time, and Special Relativity was born. Heisenberg applied a similarly rigorous approach to atomic phenomena and discovered quantum mechanics. Once Mach's ideas did their job, however, they could be cast away as nonsense."

It was time for Julie and Peter to go. After we waved good-bye, I returned to my computer. Sitting in front of the screen, I kept thinking about Peter's question. Julie's response, I thought, might explain much more than Mach's influence at the turn of the century. The need to be informed by patently wrong ideas may have been operative on a larger scale as well. The present time is clearly a time of a major paradigm shift, but consider the last great paradigm shift, the one that resulted in the Newtonian framework. This previous paradigm shift was profoundly problematic for the human spirit. It led to the conviction that we are strangers, freaks of nature, conscious beings in a universe that is almost entirely unconscious, and that, since the universe is strictly deterministic, even the free will we feel in regard to the movements of our bodies is an illusion. Yet it was probably necessary for the Western mind to go through the acceptance of such a paradigm.

The overwhelming success of Newtonian physics led most scientists and most philosophers of the Enlightenment to rely on it exclusively. As far as the quest for knowledge about reality was concerned, they regarded all of the other modes of expressing human experiences, such as accounts of numinous experiences, poetry, art, and so on, as irrelevant. This reliance on science as the only way to the truth about the universe is clearly obsolete. Science has to give up the illusion of its self-sufficiency and the self-sufficiency of human reason. It needs to unite with other modes of knowing, in particular with contemplation, and help each of us move to higher levels of being and toward the Experience of Oneness.

If this is indeed the direction of the emerging world-view, then the paradigm shift we are presently going through will prove to be nourishing to the human spirit and in correspondence with its deepest conscious or unconscious yearning—the yearning to emerge out of Plato's shadows and into the light of numinosity.

APPENDICES

The Relativity of Simultaneity and the Relativity of Length

1. Proof of the Relativity of Simultaneity

The full story of Julie and Peter's encounter will establish the relativity of simultaneity as an unavoidable consequence of the two basic postulates of Special Relativity.

To make the point of the story as clearly as possible, we will endow Peter and Julie with superhuman capacities: Their reaction times are billions of times better than those of ordinary mortals; if they receive signals, or other impressions, that differ in their arrival times even by no more than a billionth of a second, they can tell them apart. We have to endow our protagonists with such superhuman capacities because some of the events in the story will unfold in very quick succession, due to the enormity of the speed of light. So, here is the full story.

Once upon a time Peter and Julie were traveling in their respective spaceships way out in space, far away from stars and planets. The spaceships were made of strong, transparent plastic. They had clocks at each end; each pair of clocks was synchronized: Peter and Julie, standing in the middle of their vehicles, acted as "midway observers."

The two spaceships approached each other. They traveled in precisely opposite directions, and, because of Peter's carelessness, they almost collided head-on. Fortunately, the spaceships just missed each other as they zoomed past. The encounter, while not disastrous, was not uneventful. This is how Julie told me about it (see Figure 2.1):

"I was just standing there, in the middle of my spaceship, which was at rest, when Peter's spaceship appeared, zooming toward me at an enormous speed. I was relieved when I realized that it had not hit my spaceship, but right at the instant his spaceship was side by side with mine, a strange incident happened: Two pieces of space debris, two small rocks, hit our vehicles. One of them hit the front end of my spacecraft, and simultaneously the rear end of Peter's (the front end of my spaceship and the rear end of his were almost touching at this point), while the other piece hit, simultaneously, the rear end of my spaceship and the front end of his (these parts of our spaceships were also practically touching). The rocks were not big enough to penetrate either Peter's ship or mine; they were big enough, however, to scratch our spaceships, and, moreover, the jolts were so violent that my two clocks stopped. Later Peter told me that his clocks stopped as well. These two jolts happened at the same time."

"How do you know that they happened at the same time?" I asked.

"Well," Julie responded, "I was standing in the middle of my spaceship, and I saw the two rocks hitting at the same time—you remember that our spaceships were transparent, so I could see what was going on both inside and outside my vehicle. Moreover, my two clocks, which stopped when jolted, showed 12:00, and since they are still broken, they still do. By the way," she added, "I didn't *see* these rocks hit at 12:00, but slightly later, because it took a little time for the light to travel from the two ends of the vehicle to the middle, where I was."

"Just out of curiosity," I said, "let me ask you this: A bird sees in two opposite directions at the same time because of the way its eyes are placed in its head. But you are not a bird. How did you see simultaneously the two rocks hitting the opposite ends of your spaceship?"

Julie eyed me suspiciously. "I was actually looking straight ahead," she responded. "I was looking at two small mirrors that were right in front of my face. They were oriented in such a way that I could see both ends of my spaceship simultaneously, at a glance, by looking at them. And I did notice that it was precisely 12:00 when Peter, who was standing in the middle of his spaceship, zoomed past me. At precisely 12:00 our spaceships were exactly aligned."

"And how do you know that?" I asked.

"There is a third clock, a small one, hanging right in the middle of my spacecraft, under the mirrors; when Peter was opposite me, it showed 12:00. I synchronized this third clock with the other two clocks that same morning."

"But, you know," I said, "Peter didn't see the two events at the same time. He told me that he saw the rock hitting the rear end of your spacecraft and the front end of his (Let's call it event A) before he saw the rock hitting the front end of your vehicle and the rear end of his (event B)."

Julie became thoughtful, but not for long. "Look," she said, "at 12:00 exactly, when we zoomed past each other, neither of us saw either event A or event B yet; remember, the events *happened* at 12:00, and light from the ends of our spaceships needed some time to reach us. By the time I saw these two rocks hitting, Peter wasn't opposite me anymore. He was closer to the rear end of my vehicle than to the front, so the light traveling from event A reached him before the light from event B did. He was closer to where event A happened, so the light from event A had less distance to travel. That's why he saw it first. But the two events really happened simultaneously, at 12:00."

"Sorry," I replied, "that's not how he explained what happened. To begin with, he claimed that *he* was at rest, and *your* spaceship was in motion, approaching his at this enormous speed. As you remember from your studies of the first postulate of Special Relativity, his claim of being at rest is just as valid as yours. He said, 'I was standing in the middle of my spaceship when the two rocks hit. I saw event A first and event B a little later. Since I was standing right smack midway between them, and my spaceship was at rest, I must conclude that event A really happened first, event B a little later.'"

This time Julie was thoughtful for a long time. Finally she said: "What about the broken clocks at the front and rear of his vehicle? They stopped when the rocks hit. What do they read?"

When I answered, she could hardly believe her ears: "*The clock in the front of his spaceship shows a fraction of a second before 12:00; the one at the rear, a fraction of a second after. According to his clocks, at 12:00 event A was already in the past, while event B hadn't happened yet.*"

As our story shows, the relativity of simultaneity follows from the two basic postulates of Special Relativity. The collections of events that comprise "the world now" are different for two frames of reference that are moving relative to each other.

2. The Relativity of Length

We will explain length contraction and its relationship with the relativity of simultaneity by continuing the story of my conversation with Julie.

When I told Julie that the two events, A and B, which were simultaneous according to her clocks were not simultaneous according to Peter's, she was shocked. She kept going back to the arguments that led us to this conclusion, trying to discover some error. As I watched her, I noticed that she kept looking at the two drawings I made, Figure 2.1 and Figure 2.2. Finally she turned to me and said, "Look at these drawings. Something is terribly wrong here. They contradict each other; they can't both be right!"

"Why not?"

"Because in Figure 2.1 my spaceship is as long as Peter's, while in Figure 2.2 it is shorter than Peter's. Our two spaceships cannot be equal in length and also unequal in length! Why did you draw my spaceship in Figure 2.2 the way you did?"

"I didn't have any choice. Look, Julie. Figure 2.2 represents a snapshot of the situation at 12:00 according to Peter's clocks (let's call it '12:00 Peter's time'). At that time event A was already in the past. The rock that hit the rear of your spaceship and the front of his had already done the damage. This means that by 12:00 Peter's time the rear of your vehicle had already zoomed by the front of his. And since, in the drawing, your vehicle is moving to the left, I drew the rear of your spaceship to the left of the front of his."

"I see," Julie responded. "And I also understand why you drew the front of my spaceship to the right of the rear of his. This is because according to his clocks event B hasn't happened yet."

"Exactly."

"But how can my vehicle be as long as Peter's and also shorter than Peter's?" Julie was getting exasperated.

"When the two vehicles are standing at rest, in the hangar, side by side, are they of equal length, or is your vehicle shorter?" I asked.

"Well, actually, I just happened to be at the hangar this morning," Julie responded. "Our two spaceships are, in fact, parked next to each other. I noticed that my spaceship is shorter—although not by as much as Figure 2.2 shows."

"Then how do you account for Figure 2.1?" I asked. "You know that Figure 2.1 is accurate. You saw the front of your spaceship right next to the rear of his at the same time that you saw the rear end of your spaceship right next to the front of his!"

Julie was baffled. "I see these two contradictory pictures in front of my mind's eye," she said after a while. "I see our two spaceships perfectly aligned, for an instant, as Peter's ship zooms past mine with the two rocks hitting; I remember seeing this perfect alignment as I looked at the two little mirrors that are hanging in the middle of my ship. And I also see myself at the hangar, just a few hours ago, looking at these two spaceships parked next to each other, mine shorter than Peter's. What's going on?!"

I kept quiet. I had learned through a long teaching career that, if at all possible, it is better to let people figure things out for themselves. I also knew how bright Julie was. She kept at it for a long time. When she finally muttered to herself, "That will do it, but it isn't possible!" I knew that she got it. Turning toward me, she said, "Is it possible that if I measure the length of an object that is moving relative to me, and I measure the length of the same object when it is at rest next to me, I get different answers?"

"Let's look at it," I answered. "We know how to measure a spaceship at rest. Just put a bunch of yardsticks next to it and see how many it takes to cover the length of the ship. But how do you measure the length of a spaceship as it zooms past you?"

It didn't take Julie long to figure it out. "Take pictures," she said. "Place a line of many small cameras, practically touching each other, all at rest in *your* spaceship. Connect them to a computer and program all of them to click at precisely 12:00 your time. See which camera took the picture of the front end of the passing spaceship and which camera took the picture of the rear end. The distance between these two cameras is the length of the passing vehicle."

"Good," I said. "The result of your measurement depends, critically, on all the cameras shooting at precisely the same time, right?"

"Of course. If they shoot at different times, the distance between the cameras will not represent the length of the spaceship." Julie stopped. An instant later she broke into a broad smile. "Of course," she said. "How could I be so stupid! Since simultaneity is relative, what is the same time for me is different times for Peter. No wonder the length of my spaceship as measured by his cameras is different from the length of my spaceship as measured by mine."

"Yes," I said. "In addition to the relativity of simultaneity we have to get used to the relativity of length. The result of length measurement of a moving object is different from the length of the same object measured at rest. And, as you have just found out, the two relativities, that of simultaneity and that of length, are interconnected; and there are other quantities, such as duration, which we tend to think of as absolute, the same for everybody, and which prove to be relative. Special Relativity brought about quite a revolution in our understanding of space and time."

~APPENDIX 2
The Proof of Bell's Inequalities

The proofs of Clarissa Gill's inequality, as well as Bell's inequalities, are straightforward. To present them we will proceed as follows: First, we introduce certain symbols for quantities that feature in the proof of Gill's inequalities; then we will prove one of her inequalities (it is enough to prove one; the proofs of the others are analogous) and conclude by proving one of Bell's inequalities.

1. Symbols

Let us denote by $A+$ a positive (yes) response to question (a), by $A-$ a negative (no) response to the same question, by $B+$ a positive to question (b), and so on. Furthermore, $n(A+)$ will denote the number of wives who are predisposed to answer yes to question (a), $n(A-)$ the number of wives who are predisposed to answer no to question (a), etc. (Since the husbands' predispositions are always contrary to the wives', it is enough to count the wives' predispositions.) The term $n(A+, B+)$ is the number of wives who are predisposed to answer yes to both questions, (a) and (b); corresponding meanings are assigned to $n(A+, B-)$, $n(A+, C+)$, etc. Finally, $n(A+, B+, C+)$ is the number of wives who are predisposed to answer yes to all three questions, with corresponding meanings assigned to $n(A+, B+, C-)$, $n(A-, B-, C+)$ etc.

The symbols we have defined so far refer to *predispositions*. Predispositions, however, cannot be measured. Recorded in the experiment are the simultaneous answers of husbands and wives to the questions posed to them. To process these records we need a new set of symbols: Let $n[A+, B+]$ (using square brackets instead of the round ones we used before) denote the number of cases in which the wife answered yes to question (a) and the husband answered yes to question (b), $n[A+, C-]$ denote the number of cases in which the wife answered yes to question (a) and the husband answered no to question (c), and so on.

Clearly, our new symbol, $n[A+, B+]$, is closely related to the one we used before, $n(A+, B-)$, because we are dealing with contrary couples: "Husband answers yes to question (b)" implies "wife has a predisposition to answer no to question (b)."

2. Proof of a Gill's Inequality

Let us begin with predispositions and prove that the quantity $n(A+, B+)$ is smaller than or equal to (but can never be greater than) the sum of the quantities $n(A+, C+)$ and $n(B+, C-)$. The proof proceeds in two steps.

First step: Since those wives who are predisposed to answer yes to all three questions are obviously predisposed to answer yes to questions (a) and (c), $n(A+, B+, C+)$ is less than or equal to (and could not possibly be more than) $n(A+, C+)$. Similarly, $n(A+, B+, C-)$ is less than or equal to (and could not possibly be more than) $n(B+, C-)$.

Second step: Since $n(A+, B+)$ is the number of wives who are predisposed to give positive responses to the first two questions, regardless of their response to the third, $n(A+, B+)$ must be equal to the sum of $n(A+, B+, C+)$ and $n(A+, B+, C-)$.

Now, according to the first step, the sum of $n(A+, B+, C+)$ and $n(A+, B+, C-)$ is smaller than or equal to (but never greater than) the sum of the quantities $n(A+, C+)$ and $n(B+, C-)$. And according to the second step, this sum of $n(A+, B+, C+)$ and $n(A+, B+, C-)$ is equal to $n(A+, B+)$. It follows that $n(A+, B+)$ is smaller than or equal to (but never greater than) the sum of the quantities $n(A+, C+)$ and $n(B+, C-)$, which is what we set out to prove. This is a Gill's inequality about predispositions.

Having proved an inequality that refers to predispositions, we can proceed to prove the analogous inequality that refers to actual answers. As stated above, because we are dealing with contrary couples, the symbol $n[A+, B+]$ is closely related to the symbol $n(A+, B-)$: "Husband answers yes to question (b)" implies "wife would have answered no to question (b)." Similarly, $n[A+, C-]$ is closely related to $n(A+, C+)$, and $n[B+, C+]$ is closely related to $n(B+, C-)$. Therefore Gill's inequality about predispositions corresponds to the following Gill's inequality about the record of experiments:

$n[A+, B-]$ is smaller than, or equal to (but never greater than) the sum of the quantities $n[A+, C-]$ and $n[B+, C+]$.

3. The Proof of Bell's Inequalities

To formulate and prove Bell's inequalities we need new symbols, symbols that refer to pairs of particles instead of contrary couples. Let us use the following notation: Any time a random decision-maker chooses to measure the spin component in the direction of a and the result of the measurement is "up," we denote this measurement, and its result, by the symbol $a+$. If a is chosen and the result is "down," the corresponding symbol is $a-$. If b is chosen and the result is "up," the symbol is $b+$, and so on. All together we have six symbols: $a+$, $a-$, $b+$, $b-$, $c+$ and $c-$.

Now we introduce two types of N-symbols, symbols with round brackets, like $N(a+, b+)$, and symbols with square brackets, like $N[a+, b+]$. When the Bell experiment is repeated many times, $N(a+, b+)$ denotes the number of A particles with spin components "up" in both the a and b directions. This symbol, and others like it, are analogous to Clarissa Gill's $n(A+, B+)$ symbol and others like it. The second type of symbol, $N[a+, b+]$, refers to results of simultaneous measurements on both particles, A and B. It is defined as follows: We denote a measurement in which direction a was chosen by the random decision-maker for particle A, with the result of the measurement being "up", while b was chosen for particle B, with the result of measurement being also "up," by the symbol $[a+, b+]$; the first letter in the square brackets always refers to particle A, and the second letter to particle B. If the Bell experiment is repeated many times, the number of times in which direction a is chosen by the random decision-maker for particle A and the result is "up," and b is chosen for particle B with the result also "up," will be denoted by $N[a+, b+]$. Corresponding meanings apply to other similar symbols, such as $N[b-, c+]$, $N[b-, c-]$, etc. These symbols are again analogous to Clarissa Gill's symbols: $N[a+, b+]$ corresponds to Gill's $n[A+, B+]$, $N[b-, c+]$ corresponds to Gill's $n[B-, C+]$, and so on.

Bell's inequalities are relations involving the N-quantities with the square brackets. They are based on Einstein's assumption regarding the EPR set-up: Both particles, A and B, have well-defined spin components in all directions, in spite of the fact that according to quantum mechanics only one component of each particle can be measured. If we now replace the contrary couples by pairs of particles and "translate" from the psychological to the EPR context, then the same logical steps that led to "Gill's inequality" will lead us to Bell's inequalities, a typical example of which is the following:

The quantity $N[a+ b-]$ is smaller than or equal to (but never greater than) the sum of the quantities $N[a+ c-]$ and $N[b+ c+]$.

That is, if you add up the two numbers $N[a+ c-]$ and $N[b+ c+]$, you come up with a number that is either greater than $N[a+ b-]$ or equal to it. The necessary "translation" is this: Replace questions (a), (b), and (c) in the story of the contrary couples by spin components along the a, b and c directions; replace a positive response in the story with a measurement that yields spin "up" and a negative response with a measurement that yields spin "down"; and, finally, replace the assumption that husbands and wives are predisposed to have definite answers to all three questions by the assumption that particles A and B definitely have "up" or "down" spin components in the directions of all three axes, a, b, and c. Once this "translation" is complete, the proof of Gill's inequality "translates" to a proof of the Bell's inequality.

~APPENDIX 3
The Quantum Theory: Interpretations and Modifications

This appendix contains material pertaining to various interpretations and modifications of the mathematical formalism of quantum mechanics, material whose inclusion in the main text would have disrupted the flow of ideas.

The appendix contains five sections. The first section explains the sources and origin of the approach to the interpretation of quantum mechanics which is presented in this book. This approach is based on Heisenberg's mainstream interpretation. It contains, however, two novel features: a new approach to the understanding of the collapse of quantum states and a resolution of the dichotomy of the ontic and epistemic interpretations of quantum states.

Other interpretations of quantum mechanics are presented in Sections 2 through 5. A comprehensive review of all the interpretations that have been suggested since the 1920s is outside the scope of this book. This appendix contains presentations of those interpretations that attracted the most attention in both professional and popular circles, and critiques of these interpretations.

1. The Present Approach: A Mainstream Interpretation with Novel Features

The interpretation of quantum mechanics presented in Chapter 4 is Heisenberg's mainstream interpretation. Its main feature is the introduction of the concept of "fields of potentiality" as a new category of existence, one that is essential to the understanding of the nature of quantum waves. As we saw in Chapters 10 and 11, this interpretation calls for the introduction of the concept of "the collapse of quantum states," a concept which expresses the transition from the potential to the actual. This concept turned out to be problematic (see Sections 2 through 5 below). I believe that my novel approach to the understanding of the collapse, the approach that is presented in Chapters 10 and 11, resolves the difficulties.

As the narrative in Chapter 11 indicates, the original impetus for the development of the present interpretation of the collapse of quantum states can be traced to a conversation I had with Paul Dirac back in 1976. In fact, I consider my interpretation of the collapse to be, for the most part, an elucidation, articulation, and elaboration of Dirac's position. Of course, there is no way to know whether Dirac would have endorsed it. This interpretation was published in 1993 in a paper entitled "The Collapse of Quantum States: A New Interpretation."[1]

The development of this understanding of the collapse was preceded by a number of papers that were devoted to the issue of the meaning and significance of quantum states.[2] In these papers the issue of the ontic versus epistemic interpretation of quantum states is resolved. It is shown that difficulties inherent in either the ontic or the epistemic interpretation disappear when the quantum state of a quantum system is understood as representing *available* or *potential knowledge* about the system. This idea is explained in Section 8 of Chapter 16. The series of papers mentioned above contains developments of this idea in both the relativistic and non-relativistic contexts, developments that are tangential to the main line of this book and have not been included in it.

The idea that quantum states represent available knowledge is in line with a remark attributed to Bohr: "It is wrong to think that the task of physics is to find out how nature is. Physics concerns what we can say about nature."[3]

2. The Many-Worlds Interpretation

The origin of the many-worlds interpretation goes back to a paper published by Hugh Everett III in 1957.[4] Everett's paper deals with many states of mind; these were later "translated" into many universes. The proponents of this interpretation claim to solve the riddle of the collapse of quantum states by postulating that the collapse does not occur at all! According to them, instead of leading to a collapse, an act of measurement leads to actualization of all the results it could possibly yield, because the universe in which the measurement was carried out splits into as many universes as the number of possible outcomes, and each of these possible outcomes is actualized in one of them.[5]

Suppose, for example, that a measurement of the component of the spin of an electron in some direction is being prepared. The outcome would be recorded as a dot that would appear at one of two possible locations on a photographic plate; One location corresponds to spin "up," the other to spin "down." According to the mainstream interpretation, whenever such a measurement is carried out, the state of the electron would collapse into either "up" or "down," a collapse that would lead to the appearance of a dot in one location or another. Not so, say the proponents of the many-worlds interpretation. The whole universe (they claim) splits into two universes. In one of them (call it "universe 1") the result is "up"; in the other ("universe 2") the result is "down." If you observe the result as "up," this simply means that you happen to inhabit universe 1. The other you, the one who inhabits universe 2, is sure to observe the result as "down."

The many-worlds interpretation is an audacious proposal. At first glance it seems as if it cannot be either confirmed or refuted: Since, presumably, there is no communication among the different universes, the only universe whose existence is verified is the one we find ourselves in. The existence of the others cannot be proved or disproved.

A closer look at this proposal reveals serious weaknesses. First, the universes that are supposedly created in a split are not always well defined. The above-mentioned example of a spin-component measurement is straightforward. But what about a measurement which determines, approximately, the position of an electron? The number of possible quantum states that can arise from such a measurement is infinite! Additionally, the process of the splitting of a universe does not jibe with Special Relativity: The process is supposed to be instantaneous, but what is instantaneous in one frame of reference is not instantaneous in another.

Furthermore, it has been pointed out that the many-worlds proposal does not really resolve the ambiguities related to the collapse; it merely translates them into a new context. The question "At what stage in the measurement process does the collapse occur?" gets translated into the question "At what stage in the measurement process does the universe split?"

There are additional arguments against the many-worlds interpretation, but we can stop here. This proposal has a certain appeal to the popular imagination, but it is hardly a satisfactory alternative to the idea of collapse.

3. Does Human Consciousness Bring the Collapse About?

The question "At what stage in the measurement process does the collapse occur?" led a small but venerable minority of quantum physicists to suggest that the collapse occurs when a human being becomes conscious of the result of a measurement, not before. This idea was first considered by J. von Neumann in his 1932 book on the mathematical foundations of quantum mechanics. It was elaborated by F. London and E. Bauer in 1939; even E. Wigner was sympathetic to it.[6]

Why would anyone wish to involve human consciousness in the collapse of quantum states? Here is one way to present the argument: Since the material world is composed of quantum systems, any material object, no matter how complex, should be describable, in principle, by the equations of quantum mechanics. These equations do not include a collapse. In a sense, the collapse is extraneous to the mathematical formalism of the quantum theory. Hence it stands to reason that the place of its appearance would be extraneous to the world of objects. But the world of objects is not the whole of reality; the consciousness of the observer is, indeed, extraneous to it. Therefore it is natural to assume that the meeting place of consciousness and matter, i.e., the act of conscious observation, is the place where the collapse is applied to the material system.

This is an elegant point of view; it has the advantage of logical consistency. Unfortunately, it does not survive close scrutiny. A series of objections to it was raised by A. Shimony in 1963.[7] Here I would like to present my own argument against this idea.

Suppose we accept the idea that the collapse occurs when a human being becomes aware of the result of an experiment. Consider the following scenario. A measurement of the spin component of an electron in some direction (call it the z-direction) is carried out automatically, i.e., without the presence of human beings, and the result appears on the printout of a computer. The printout is a superposition of two statements: (1) "The z-component of the spin is up." (2) "The z-component of the spin is down." If it is human consciousness that collapse the quantum state, then, presumably, the superposition collapses when someone reads the printout and not before.

Now assume the following: The ink cartridge of the printer has just enough ink to print statement 1, and not enough ink to print the slightly longer statement 2. So the red light indicating the need to replace the cartridge would appear if and only if statement 2 is being printed. Furthermore, assume that the appearance of the red light is accompanied by the sound of a bell. If the bell sounds, a secretary, sitting in an adjacent room, would hear it and proceed to change the cartridge.

Now consider the two possibilities: First, the secretary does hear the bell. In this case, clearly, the collapse occurs first, and the involvement of human consciousness with the result of the experiment occurs later. The other possibility, namely, the case in which the bell is not heard, is even more interesting: In this case there is a collapse without any human involvement with the result of the experiment.

4. Bohm's Causal Interpretation

David Bohm's causal interpretation of quantum mechanics is remarkable on two counts. First, it disproved von Neumann's theorem concerning hidden variable theories (see Chapter 7, Section 2), and, second, it provided an interpretation of quantum mechanics that is both causal and realistic.

Bohm's theory can be considered an elaboration of an idea that was suggested by L. de Broglie at the 1927 Solvay Conference. Bohm, however, was unaware of de Broglie's suggestion when he developed his theory. Bohm published his interpretation in 1952 and continued to develop it, in collaboration with V. Vigier, B. Hiley, and others, till his death in 1992.[8]

Bohm's theory is mathematically equivalent to quantum mechanics. He arrived at it by carrying out certain mathematical manipulations on Schrödinger's equation. These manipulations led to a new interpretation of the wave-particle duality. In the standard interpretation the waves represent potentialities; a particle appears as a result of a measurement. In Bohm's interpretation the concept of potentiality is dispensed with; the particle and the wave are both actual. The wave is a kind of "pilot wave," guiding the particle by the action of a minute "quantum force," a force that is generated by "a quantum potential." Each particle is actual at all times, following a precise trajectory, one which is completely determined by the quantum potential through the pilot waves it generates. That is why Bohm's interpretation is considered causal: Indeterminism and randomness are gone!

Why, then, can *we* not predict these trajectories precisely? Well, each trajectory is determined by the initial conditions, i.e., the precise values of the position and the velocity (or momentum) at a given time. But the uncertainty principle prevents us from knowing these initial conditions. Since we do not know the initial conditions precisely, we cannot calculate the precise trajectories.

Thus, according to Bohm, the uncertainty principle is epistemic rather than ontic. It limits what we can know; it does not limit the properties of particles. At each moment each particle has a precise position and a precise momentum. These precise values are unknowable; they are "hidden variables."

As mentioned above, Bohm's interpretation is a remarkable accomplishment. It was achieved, however, at a high price. Consider the following weaknesses:

(1) As D. Wick points out,[9] the probability law of the particle's location is involved in the calculation of the quantum force acting on it. Probabilities have to do with ensembles; why should they be involved in the calculation of the trajectory of a single particle?

(2) Calculation of the velocity of the electron in the ground state of the hydrogen atom yields zero. The electron is not moving at all! This result does not contradict anything, but it is counterintuitive.

(3) Locality is violated. The communication between the quantum potential and the particle occurs faster than the speed of light; it is supposed to be instantaneous. The violation of locality is, of course, an important aspect of Bell's correlations. In the context of the mainstream interpretation of quantum mechanics, however, this violation has to do with the transition from the potential to the actual (Chapter 16, Section 9) and cannot be used to transmit signals (Chapter 7, Section 7). In Bohm's interpretation locality is violated in the transmission of information between two actual entities–a clear violation of Einstein's Special Theory of Relativity.

(4) Bohm's interpretation cannot be reconciled with the principle of relativity: In the relativistic context it cannot be valid in all inertial frames of reference. It can be true only in one "preferred" frame. This has to do with the relativity of simultaneity (Chapter 2, Section 6). The communication between the particle and the pilot wave is supposed to be instantaneous. But, as we mentioned before, that which is instantaneous in one inertial frame of reference is not instantaneous in another.

5. The GRW Approach

In 1986 G. C. Ghiraldi, A. Rimini, and T. Weber (GRW) published a paper in which they proposed to solve the problem of the collapse by modifying quantum mechanics.[10] In the modified version the collapse takes place as an integral part of the theory.

In GRW's theory the quantum state of a quantum system develops according to Schrödinger's equation. At certain randomly selected instants, however, this development is arrested and the quantum state spontaneously collapses into a well-localized state. For a single particle the probability of such a spontaneous collapse is so low that, in practical terms, the predictions of the theory are the same as those of quantum mechanics. But for a macroscopic system, i.e., a system consisting of a very large number of particles, this spontaneous collapse becomes a rather frequent event. This explains why whenever we observe macroscopic systems we see them in a collapsed state.

The attractive feature of the theory is the inclusion of the collapse as an integral part of its formalism. The theory has its weaknesses, however. First, according to the uncertainty principle, when the state becomes well localized in space, its momentum becomes more uncertain; this leads to violations of the laws of conservation of momentum and energy. Second, this theory, just like Bohm's interpretation, is incompatible with the principle of relativity: The spontaneous collapse occurs instantaneously, and simultaneity is relative to the frame of reference. A collapse that is instantaneous in one inertial frame is not instantaneous in another. Therefore the theory must apply in one "preferred" frame of reference only. The principle of relativity, which demands that the theory apply to all inertial frames, is violated. Third, as Albert and Vaidman point out,[11] the collapse in GRW's theory is a collapse into a well-localized position, but some measurements require a different kind of collapse, e.g., a collapse into a well-defined energy state.

In 1990 P. Pearle joined forces with G. C. Ghiraldi and A. Rimini and proposed an alternative theory, which involves a continuous localization process instead of GRW's random collapses.[12] While this theory may be amenable to a relativistic generalization, it does not resolve the objection raised by Albert and Vaidman.

Finally, taking a broad perspective on these two theories (the original GRW theory and the one that involves continuous localization) as well as other theories that propose modifications of quantum mechanics, I agree with a recent statement by Anton Zeilinger:

Suggestions actually to change quantum mechanics are not just interpretations but are really alternative theories. In view of the extremely high precision with which the theory [i.e., quantum mechanics] has been experimentally confirmed, and in view of its superb mathematical beauty and symmetry, I consider a final success of such attempts extremely unlikely.[13]

NOTES

Chapter 1

Epigraph: A. Einstein, "Autobiographical Notes," in P. A. Schilpp, ed., *Albert Einstein: Philosopher-Scientist*, p. 21.

1. W. Heisenberg, *Physics and Beyond*, p. 62.
2. Ibid., p. 61.
3. Ibid., p. 60.
4. Ibid., p. 63.
5. Ibid.
6. Ibid., p. 64.

Chapter 2

Epigraph: T. S. Eliot, *Four Quartets: Burnt Norton*, p. 13.

1. A. Pais, "*Subtle Is the Lord*," p. 5.
2. Lucretius, *On the Nature of the Universe*, p. 39
3. I. Newton, quoted in G. Holton, *Introduction to Concepts and Theories in Physical Science*, p. 298.
4. St. Augustine, *Confessions* XI.14, p. 264; XI.28, p. 277; XI.27, p. 276.
5. A. N. Whitehead, *Science and the Modern World*, p. 118.
6. St. Augustine, *Confessions* XI.12, p. 262.
7. Ibid., XI.14, p. 263.
8. Ibid., XI.13, p. 263.
9. A. Calaprice, collector and ed., *Quotable Einstein*, p. 61.
10. There seems to be some indirect evidence that toward the end of his life Einstein softened his opposition to indeterminism. A letter that W. Pauli wrote to M. Born in 1954 contains the following passage (quoted in M. Jammer, *Einstein and Religion*, p. 53): "Einstein does not consider the concept of 'determinism' to be as fundamental as it is frequently held to be (as he told me emphatically many times). . . . [He] disputes that he uses as a criterion for the admissibility of a theory the question: 'Is it rigorously deterministic?'" To the best of my knowledge, however, there is no hint of the possible acceptance of indeterminism in Einstein's writings or public addresses. Furthermore, A. Fine analyzed this statement of Pauli's in detail and reached the conclusion that Pauli must have misinterpreted Einstein's position (*Shaky Game*, pp. 101–103).

Chapter 3

Epigraph: Quoted in K. Freeman, *Ancilla to the Pre-Socratic Philosophers*, p. 25.

1. H. Folse, *Philosophy of Niels Bohr*, p. 107.
2. A. Einstein in P. A. Schilpp, ed., *Albert Einstein: Philosopher-Scientist*, p. 674.

3. W. Heisenberg, *Physics and Beyond*, p. 77.

4. Ibid., pp. 77–78.

5. Quoted in H. Folse, *Philosophy of Niels Bohr*, p. 81.

6. M. C. Nahm, *Selections from Early Greek Philosophy*, pp. 38–39.

7. Quoted by A. Peterson in H. Folse, *Philosophy of Niels Bohr*, 8.

8. Ibid., 12.

9. W. Heisenberg, *Physics and Philosophy*, Chapter 3. Generally speaking, Heisenberg's position on the issue of the epistemic versus the ontic interpretation of quantum mechanics is complex. We will discuss it in Section 5 of Chapter 4.

Chapter 4

Epigraph: W. Shakespeare, *Macbeth*, Act I, Scene 3.

1. W. Moore, *Schrödinger*, p. 192.

2. W. Heisenberg, *Physics and Beyond*, pp. 73–75.

3. Ibid.

4. Aristotle, *Selections*, p. 71.

5. The ontic interpretation presented in this chapter is called "Heisenberg's" because most physicists learned it from Heisenberg's writings. Shimony points out (*Search for a Naturalistic World View* vol. 2, p. 313), however, that the idea that quantum states represent potentialities was first proposed by H. Morgenau, who used the equivalent term "latency." (*Philosophy of Science* 16 (1949), p. 287.)

6. W. Heisenberg, *Physics and Philosophy*, p. 53.

7. W. Heisenberg, *Daedalus* 87 (1958), p. 99.

Chapter 5

Epigraph: Quoted in B. N. Kursunoglu and E. P. Wigner, *Paul Adrien Maurice Dirac*, p. 148.

1. W. Heisenberg, *Physics and Beyond*, p. viii.

2. Quoted by A. Pais in B. N. Kursunoglu and E. P. Wigner, *Paul Adrien Maurice Dirac*, p. 93.

3. Quoted by L. M. Brown and H. Rechenberg in ibid., p. 136.

Chapter 6

Epigraph: A. Einstein in a letter to M. Born, December 4, 1926, quoted in G. Holton, *Thematic Origins of Scientific Thought: Kepler to Einstein*, p. 120.

1. A. Pais, "*Subtle Is the Lord*," pp. 6–7.

2. Ibid., p. 6.

3. Ibid., p. 445.

4. W. Heisenberg, *Physics and Beyond*, p. 80.

5. L. Rosenfeld quoted in J. A. Wheeler and W. H. Zurek, eds., *Quantum Theory and Measurement*, p. ix.

6. A. Pais, "*Subtle Is the Lord*," p. 449.

7. A. Einstein, B. Podolsky, and N. Rosen, "Can Quantum-Mechanical Description of Physical Reality Be Considered Complete?" *Physical Review* 47 (1935), p. 777.

8. Quoted by N. D. Mermin in A. P. French and P. J. Kennedy, eds., *Niels Bohr: A Centenary Volume*, p. 142.

9. N. Rosen in P. C. Aichelberg and R. U. Sexl, eds., *Albert Einstein: His Influence on Physics, Philosophy, and Politics*, p. 66.

Chapter 7

Epigraph: A. Calaprice, collector and ed., *Quotable Einstein*, p. 170.

1. J. S. Bell, *Speakable and Unspeakable in Quantum Mechanics*, p. 139.
2. Interview, *Omni*, May 1988.
3. B. d'Espagnat, *Scientific American* 241 (November 79), p. 158. The writing of the next section was inspired by this article.
4. A. Aspect, P. Grangier, and G. Roger, *Physical Review Letters* 47 (1981), p. 460.
5. See N. D. Mermin, *Physics Today*, June 1990, p. 9.
6. L. Hardy, *Physics Letters A* 167 (1992), p. 17.
7. C. H. Bennett, G. Brassard, C. Crépou, R. Jozsx, A. Peres, and W. K. Wooters, *Physical Review Letters* 70 (1993), p. 1895. D. Bouwmeester, J.-W. Pan, K. Mattle, M. Eibl, H. Weinfurter, and A. Zeilinger, *Nature* 390 (1997), p. 575. D. Boschi, S. Branca, F. De Martini, L. Hargy, and S. Popescu, *Physical Review Letters* 80 (1998), p. 1121.
8. N. Rosen in P. C. Aichelberg and R. U. Sexl, eds., *Albert Einstein, His Influence on Physics, Philosophy, and Politics*, p. 67.

Chapter 8

Epigraph: A. N. Whitehead, *Science and the Modern World*, p. 51

1. Ibid., pp. 70–71.

Chapter 9

Epigraph: E. Schrödinger, *What Is Life?* with *Mind and Matter* and *Autobiographical Sketches*, p. 123.

1. Ibid., p. 118.
2. Ibid., p. 119.
3. Ibid., p. 122.
4. Ibid., p. 127.

Chapter 10

Epigraph: A. N. Whitehead, *Process and Reality*, p. 40.

1. W. Heisenberg, *Physics and Beyond*, p. 7.
2. Ibid., p. 8.
3. Ibid.
4. Ibid., p. 9.
5. Ibid., p. 10.
6. Ibid., pp. 10–11.
7. Plato, *Collected Dialogues*, p. 1161.
8. Ibid., p. 1165.
9. Ibid., p. 1167.
10. A. N. Whitehead, *Process and Reality*, p. 39.
11. W. Heisenberg, *Physics and Beyond*, p. 8.
12. St. Augustine, *Confessions* III.7, p. 62.
13. Plato, *Collected Dialogues*, p. 1161.

14. E. Schrödinger, *What Is Life?* with *Mind and Matter and Autobiographical Sketches*, p. 127.

15. Plato, *Collected Dialogues*, p. 529.

16. A. Pais, *Niels Bohr's Times: In Physics, Philosophy, and Policy*, p. 4.

Chapter 11

Epigraph: Plotinus, *Enneads*, III.8.1, trans. by J. N. Deck, *Nature, Contemplation, and the One*, pp. 141–42.

1. Ibid., p. 141.

Chapter 12

Epigraph: Plotinus, *Enneads*, IV.3.7, p. 303.

1. J.M. Burgers, *Experience and Conceptual Activity*, p. 113.

2. Ibid., p. iv.

3. P.D. Ouspensky, *A New Model of the Universe*, p. 321.

4. A.N. Whitehead, *Science and the Modern World*, p. 84.

5. K.W. Kelley, ed., *The Home Planet*, text with image 138.

6. L. Thomas, *Fragile Species*, p. 135.

7. E. Schrödinger, *What Is Life?* with *Mind and Matter and Autobiographical Sketches*, p. 123.

8. L. Weschler, *Shapinsky's Karma. Bogg's Bills*, p. 71.

9. E. Schrödinger, *What Is Life?* with *Mind and Matter and Autobiographical Sketches*, p. 119.

10. A. Eddington, *The Nature of the Physical World*, p. 322.

11. G.E.R. Lloyd, *From Polarity to Analogy*, pp. 233, 235, 236, 238.

12. Ibid., pp. 250–251.

13. Plato, *Timaeus*, 36d–e, in *Collected Dialogues*, p. 1166.

14. Plotinus, *Enneads*, IV.3.7, p 303.

15. W. Heisenberg, *Physics and Beyond*, p. 215.

16. E. Schrödinger, *My View of the World*, pp. 21–22.

17. A. Einstein, *The Human Side*, p. 33.

Chapter 13

Epigraph: Quoted in W. Heisenberg, *Physics and Beyond*, p. 206.

1. Quoted by G. Holton, *Introduction to Concepts and Theories in Physical Science*, p. 298.

Chapter 14

Epigraph: F. Merrell-Wolfe, *Transformations in Consciousness*, p. 163.

1. Plato, *Republic*, 510c,510e, 511a-c, in *Collected Dialogues*, pp. 745–46.

2. W. Heisenberg, *Physics and Beyond*, pp. 77–78.

Chapter 15

Epigraph: Quoted by E. Schrödinger in *Nature and the Greeks*, p. 30.

1. A. N. Whitehead, *Modes of Thought*, p. 151.

2. Ibid., p. 152.

3. M. H. Fisch, gen. ed., *Classic American Philosophers*, p. 401.

4. Ibid.

5. Ibid.
6. A. N. Whitehead, *Modes of Thought*, p. 151.
7. Ibid., p. 152.
8. A. N. Whitehead, *Process and Reality*, p. 28.
9. Ibid., p. 18.
10. A. N. Whitehead, *Modes of Thought*, p. 153.
11. Quoted by Lowe in M. H. Fisch, gen. ed., *Classic American Philosophers*, p. 407.
12. Ibid., footnote 24.
13. A. N. Whitehead, *Modes of Thought*, p. 152.
14. Ibid., p. 154.
15. Ibid., pp. 154–155.
16. Ibid., p. 156.
17. Ibid., p. 163.
18. Ibid.
19. E. Schrödinger, *Science and Humanism*, p. 17.
20. Ibid., pp. 20–21.
21. K. Wilber, ed., *Quantum Questions*, p. 51.
22. L. Carroll, *Alice in Wonderland*, p. 53.

Chapter 16
Epigraph: L. Carroll, *Alice in Wonderland*, p. 14.
1. A. N. Whitehead, *Modes of Thought*, p. 151.
2. E. Schrödinger in J. A. Wheeler and W. H. Zurek, eds., *Quantum Theory and Measurement*, p. 157.
3. This application of Whitehead's perspective is "neo-Whiteheadian" rather than Whiteheadian. For Whitehead the objectivized aspect of a single actual entity is localized in space; it cannot manifest in two locations that are far apart.

Chapter 17
Epigraph: Plotinus, *Enneads* IV.8.1, p. 410.
1. M. Proust, *Remembrance of Things Past*, vol. 3, p. 898.
2. Ibid., p. 900.
3. Ibid., p. 904.
4. Plato, *Republic and Other Works*, p. 209.
5. Plotinus, *Enneads* IV.8.1, p. 410.
6. J. N. Deck, *Nature, Contemplation, and the One*, p. 42.
7. Plotinus, *Enneads* V.1.4, p. 436.
8. J. Deck, *Nature, Contemplation, and the One*, p. 57.
9. Ibid., p. 93.
10. Plotinus, *Enneads* III.8.1, trans. J. N. Deck, *Nature, Contemplation, and the One*, p. 141.
11. Quoted in A. N. Whitehead, *Science and the Modern World*, p. 84.
12. Ibid., p. 85.
13. J. Deck, *Nature, Contemplation, and the One*, 44.
14 E. Schrödinger, *What Is Life?* with *Mind and Matter* and *Autobiographical Sketches*, pp. 118.
15. E. Schrödinger, *Science and Humanism*, pp. 20–21.

16. Quoted in K. Wilber, ed., *Quantum Questions*, p. 51.
17. A. N. Whitehead, *Process and Reality*, pp. 343–51.
18. Ibid., p. xiv.

Chapter 18

Epigraph: W. C. Segal, *Middle Ground*, p. 2.
1. J. Wheeler, "Law Without Law," in J. Wheeler and W. Zurek, eds., *Quantum Theory and Measurement*, p. 202.
2. Ibid.
3. Plotinus, *Enneads* IV.8.1–2; pp. 411–13.
4. St. Augustine, *Confessions* XI.11, p. 261.
5. C. Wilson, *Quest*, Summer 1993, pp. 19–20.
6. J. Campbell, *Power of Myth*, p. 110.
7. Plotinus, *Enneads* IV.8.1, p. 410.
8. L. Price, *Dialogues of Alfred North Whitehead*, pp. 370–71.

Chapter 19

Epigraph: A. Damiani's comments on *The Enneads* are published in Plotinus, *Enneads*, Appendix I, pp. 711–37. This quotation appears on p. 712.
1. This is Schrödinger's translation (*Science and Humanism*, p. 51). McKenna's (Plotinus, *Enneads* VI.4.14, p. 600) is "what are We?"
2. E. Schrödinger, *Science and Humanism*, p. 51.
3. Quoted in H. J. Folse, *The Philosophy of Niels Bohr*, p. 54.
4. A. Damiani, *Standing in Your Own Way*, p. 19.
5. Ibid., p. 21.
6. E, Schrödinger, *What Is Life?* with *Mind and Matter* and *Autobiographical Sketches*, p. 129.
7. Ibid., pp. 130–31.
8. Ibid., p. 130.
9. K. Brower, *Starship and the Canoe*, p. 232.
10. E. Schrödinger, *What Is Life?* with *Mind and Matter* and *Autobiographical Sketches*, p. 127.
11. Quoted in Plotinus, *Enneads*, Appendix I, p. 711.
12. Ibid., p. 708.
13. Niels Bohr Archives, "Atoms and Human Knowledge—Nicola Tesla," reprint of a lecture given at the Nicola Tesla conference, 1956; quoted in H. Folse, *Philosophy of Niels Bohr*, p. 16.
14. Ibid., p. 54.
15. Quoted in R. Tarnas, *Passion of the Western Mind*, p. 333.
16. Ibid., pp. 339–40.
17. W. James, *Principles of Psychology*, pp. 110–11.
18. Plotinus, *Enneads* VI.9.9, p. 706.
19. Ibid. VI.9.10, p. 708.
20. Ibid. VI.9.11, p. 709.
21. S. Hawking, *Brief History of Time*, p. 174.
22. W. Heisenberg, *Physics and Beyond*, p. 63.

Epilogue
1. St. Augustine, *Confessions* XI.II, p. 261.

Appendix 3
1. S. Malin, *Foundations of Physics* 23 (1993), p. 881.
2. S. Malin, *Physical Reviews D* 26 (1982), p. 1330; *Physical Reviews D* 29 (1984), p. 1856; *Foundations of Physics* 14 (1984), p. 1083; *Foundations of Physics* 16 (1986), p. 1297.
3. Quoted by A. Peterson in H. Folse, *Philosophy of Niels Bohr*, p. 8.
4. H. Everett III, *Reviews of Modern Physics* 29 (1957), p. 454.
5. See, e.g., B. DeWitt, *Physics Today* 23 (1970), p. 30.
6. J. von Neumann, *Mathematical Foundations of Quantum Mechanics*, Part VI; F. London and E. Bauer in J. Wheeler and E. Zurek, eds., *Quantum Theory and Measurement*, p. 217; "Remarks on Many-Body Problem," in E. Wigner, *Symmetries and Reflections*, p. 171.
7. A. Shimony, *Search for a Naturalistic World View*, vol. 2, p. 3.
8. D. Bohm and B. Hiley, *Undivided Universe*; F. D. Peat, *Infinite Potential: The Life and Times of David Bohm*.
9. D. Wick, *Infamous Boundary*, p. 73.
10. G. C. Ghiraldi, A. Rimini, and T. Weber, *Physical Reviews D* 34 (1986), p. 470.
11. See D. Z. Albert, *Quantum Mechanics and Experience*, pp. 100-103.
12. G. C. Ghiraldi et al., *Physical Reviews D* 42 (1990), p. 78.
13. A. Zeilinger, *Foundations of Physics* 29 (1990), p. 631.

BIBLIOGRAPHY

Aichelberg, P.C., and R. U. Sexl, eds. 1979, *Albert Einstein: His Influence on Physics Philosophy, and Politics*. Braunschweig/Wiesbaden, Germany: Vieweg.

Albert, D. Z. 1992. *Quantum Mechanics and Experience*. Cambridge: Harvard University Press.

Aristotle, 1927. *Selections*. New York: Charles Scribner's Sons. Ed. W. D. Ross.

Augustine, Saint. 1961. *The Confession*. New York: Penguin. Trans. R. S. Pine-Coffin.

Bell, J. S. !987. *Speakable and Unspeakable in Quantum Mechanics*. Cambridge: Cambridge University Press.

Bohm, D. and B. Hiley. 1993. *The Undivided Universe*. New York: Routledge.

Brower, K. 1978. *The Starship and the Canoe*. New York: Harper & Row.

Brown, L. M., and Rechenberg, H. 1987. "Paul Dirac and Werner Heisenberg—a Partnership in Science." In B. M. Kursunoglu and E. P. Wigner, eds. *Paul Adrien Maurice Dirac*. Cambridge: Cambridge University Press.

Burgers, J. M. 1965. *Experience and Conceptual Activity*. Cambridge, MA: MIT Press.

Calaprice, A., collector and ed. 1996. *The Quotable Einstein*. Princeton, N.J.: Princeton University Press.

Campbell, J. 1988. *The Power of Myth*. New York: Doubleday.

Carroll, L. 1971. *Alice in Wonderland*. New York: Norton.

Damiani, A. 1993. Standing in Your Own Way. Burdett, N.Y.: Larson.

Deck, J. N. 1991. *Nature, Contemplation, and the One*. Burdett, N.Y.: Larson.

Dirac, P. A. M. 1958. *The Principles of Quantum Mechanics*. Oxford: Clarendon.

Eddington, A. S. 1928. *The Nature of the Physical World*. New York: Macmillan.

———.1929. *Science and the Unseen World*. New York: Macmillan.

Einstein, A. 1979. *The Human Side*. Princeton, N.J.: Princeton University Press. Selected and ed. H. Dukas and B. Hoffman.

Eliot, T. S. 1971. *Four Quartets*. New York: Harcourt Brace Jovanovich.

Fine, A. 1986. *The Shaky Game*. Chicago: University of Chicago Press.

Fisch, M. H. gen. ed. 1951. *Classic American Philosophers*. New York: Appleton-Century Crofts.

Folse, H. 1988. *The Philosophy of Niels Bohr*. New York: North Holland.

Freeman, K. 1983. *Ancilla to the Pre-Socratic Philosophers*. Cambridge: Harvard University Press.

French, A. P., and P. J. Kennedy, eds. 1985. *Niels Bohr: A Century Volume*. Cambridge: Harvard University Press.

Hawking, S. 1988. *A Brief History of Time*. New York: Bantam.

Heisenberg, W. 1962. *Physics and Philosophy*. New York: Harper & Row.

———.1972. *Physics and Beyond*. New York: Harper & Row.

——.1979. *Philosophical Problems of Quantum Physics*. Woodbridge, Conn.: Ox Bow. Trans. F. C. Hayes.

Holton, G. 1973. *Introduction to Concepts and Theories in Physical Science*. 2nd ed. Reading, Mass.: Addison-Wesley.

——.1973. *Thematic Origins of Scientific Thought: Kepler to Einstein*. Cambridge: Harvard University Press.

Huxley, A. 1945. *The Perennial Philosophy*. New York and London: Harper and Brothers.

Ingram, C., C. Wilson, and M. Matousek. 1993. "*Three Views of the Millenium.*" In *Quest* 6; no. 3.

James, W. 1950. *The Principles of Psychology*. New York: Dover.

Jammer, M. 1966. *The Conceptual Development of Quantum Mechanics*. New York: McGraw-Hill.

——.1999. *Einstein and Religion*. Princeton, N.J.: Princeton University Press.

Kursunoglu, B. N., and E. P. Wigner. 1987. *Paul Adrien Maurice Dirac*. Cambridge: Cambridge University Press.

Lloyd, G. E. R. 1966. *From Polarity to Analogy*. Cambridge: Cambridge University Press.

London, F., and E. Bauer. 1939. *La Théorie de l'observation en méchanique quantique*. Paris: Hermann. In English: Wheeler and Zurek 1983, p. 217.

Lowe, V. 1951. "Introduction to Alfred North Whitehead." In M. H. Fisch, gen. ed., *Classic American Philosophers*. New York: Appleton-Century-Crofts.

Lucretius, 1982. *On the Nature of the Universe*. New York: Penguin. Trans. R. E. Latham.

Kelley, K.W., ed. 1988. *The Home Planet*. Reading, Mass.: Addison-Wesley.

Malin, S. 2004. *The Eye that Sees Itself*. Sandpoint, Idaho: Morning Light Press.

Merrell-Wolff, F. 1994. *Experience and Philosophy*. Albany: State University of New York Press.

——.1995. *Transformations in Consciousness*. Albany: State University New York Press.

Moore, W. 1989. *Schrödinger*. Cambridge: Cambridge University Press.

Nahm, M. C. 1964. *Selections from Early Greek Philosophy*. Englewood Cliffs, N.J.: Prentice-Hall.

Pais, A. 1982. "*Subtle Is the Lord*": *The Science and Life of Albert Einstein*. Oxford: Oxford University Press.

——.1991. *Niels Bohr's Times: In Physics, Philosophy and Policy*. Oxford: Clarendon.

Peat, F. D. 1996. *Infinite Potential: The Life and Times of David Bohm*. Reading, Mass.: Addison-Wesley.

Plato. 1959. *The Timaeus*. New York: Macmillan. Trans. F. M. Cornford.

——.1960. *The Republic and Other Works*. Garden City, N.Y.: Doubleday. Trans. B. Jowett.

——.1961. *The Collected Dialogues*. Princeton, N.J.: Princeton University Press. Ed. E. Hamilton and H. Cairns.

Ouspensky, P. D. 1949. *In Search of the Miraculous*. New York: Harcourt, Brace and World.

——.1967. *A New Model of the Universe*. New York: Alfred A. Knopf.

——.1968. *The Psychology of Man's Possible Evolution*. New York: Bentam Books.

Plotinus. 1992. *The Enneads*. Burdett, N.Y.: Larson. Trans. S. McKenna.

Price, L. 1954. *Dialogue of Alfred North Whitehead*. Boston: Little, Brown.

Proust, M. 1982. *Remembrance of Things Past*, vol. 3, New York: Vintage. Trans. C. K. Scott Moncrieff and Terence Kilmartin and by Andreas Mayor.

Schrödinger, E. 1951. *Science and Humanism*. Cambridge: Cambridge University Press.

——.1954. *Nature and the Greeks*. Cambridge: Cambridge University Press.

——.1958. *My View of the World*. Cambridge: Cambridge University Press. Trans. C. Hatings.

——.1992. *What is Life?* with *Mind and Matter* and *Autobiographical Sketches*. Cambridge: Cambridge University Press.

Segal, W. C. 1985. *The Middle Ground*. Brattleboro, Vt.: Stillgate.

——.1998. *Opening: Collected Writings of William Segal 1985–1997*. New York: Continuum.

Shimony, A. 1993. *Search for a Naturalistic World View*, vols. 1 and 2. Cambridge: Cambridge University Press.

Tarnas, R. 1993. *The Passion of the Western Mind*. New York: Ballantine.

Thomas, L. 1992. *The Fragile Species*. New York: Charles Scribner's Sons.

Von Neumann, J. 1932. *Mathematische Grundlagen der Quantenmechanik*. Berlin: Springer. In English: 1955. *The Mathematical Foundations of Quantum Mechanics*. Princeton, N.J.: Princeton University Press. Trans. R. T. Beyer.

Weschler, L. 1988. *Shapinsky's Karma, Bogg's Bills*. San Francisco: North Point.

Wheeler, J. A., and W. H. Zurek, eds. 1983. *Quantum Theory and Measurement*. Princeton, N.J.: Princeton University Press.

Whitehead, A. N. 1953. *Science and the Modern World*. New York: Free Press.

——.1966. *Modes of Thought*. New York: Free Press.

——.1978. *Process and Reality*. Corrected ed. New York: Free Press. Ed. D. R. Griffin and D. W. Sherburne.

Wick, D. 1995. *The Infamous Boundary*. Boston: Birkhäser.

Wigner, E. 1979. *Symmetries and Reflections*. Woodbridge, Conn.: Ox Bow.

Wilber, K., ed. 1985. *Quantum Questions*. Boston and London: Shambala.

INDEX